Cooperative Pluralism

STUDIES IN GOVERNMENT
AND PUBLIC POLICY

Cooperative Pluralism

The National Coal Policy Experiment

Andrew S. McFarland

University Press of Kansas

Published by the University Press of Kansas (Lawrence, Kansas 66049), which was organized by the Kansas Board of Regents and is operated and funded by Emporia State University, Fort Hays State University, Kansas State University, Pittsburg State University, the University of Kansas, and Wichita State University

Library of Congress Cataloging-in-Publication Data

McFarland, Andrew S., 1940–
Cooperative pluralism : the national coal policy experiment / by Andrew S. McFarland.
p. cm. — (Studies in government and public policy)
Includes bibliographical references and index.
ISBN 0-7006-0617-3 ISBN 0-7006-0618-1 (pbk.)
1. National Coal Policy Project. 2. Coal trade—Government policy—United States. 3. Coal mines and mining—Government policy—United States. 4. Energy policy—United States. I. Title. II. Series.
HD9546.M33 1993
338.2′724′0973—dc20 93-8156

British Library Cataloguing in Publication Data is available.

Printed in the United States of America.
10 9 8 7 6 5 4 3 2 1

The paper used in this publication meets the minimum requirements of the American National Standard for Permanence of Paper for Printed Library Materials Z39.48-1984.

Dedicated to Aaron Wildavsky

Who like the angel of the mines, Loretta Lynn,
at first appreciates the sentimental theme;
but then laces it with wry appraisals
of the confused culture we're living in.

Contents

Preface

When I first learned about the National Coal Policy Project, I was fascinated by the existence of an experiment in which environmentalists and business executives would attempt to negotiate agreements about controversial issues without the participation of governmental personnel. Could interest groups reach agreements about controversial regulations by negotiating among themselves, and if such agreements could be reached, would they be put into practice? As just such an experiment to deal with a broad range of issues on the national level, the National Coal Policy Project (NCPP) was a rare event in American political history.

Accordingly, I determined to write this book, and that decision fit well with my previous research about public interest groups, their lobbying techniques, and their political effectiveness. I had previously written two books in which I established that Common Cause and the environmentalist groups of the 1970s acquired considerable political effectiveness. The time had come to think of the consequences of this new form of political representation for effective public policy, including a consideration of the national need for economic efficiency in competing in international markets. The interaction of the environmentalists and the business executives in the NCPP offered an excellent forum for such an analysis.

In 1993 the NCPP seems more relevant than ever. National leaders in the fields of politics, business, scholarship, and journalism have repeatedly called for greater cooperation among normal adversaries in the interest of advancing the national economic position amid international competition. But such calls for cooperation are seldom accompanied by any suggestions about how the champions of conflicting economic interests might come to agree. The NCPP's history, on the other hand, does provide some ideas about how normally opposed groups can sometimes agree and how such agreements might be put into effect. It seems that government must be a part of this process.

This discussion of the NCPP also illuminates a number of theoretical topics of interest to political scientists and others. In the conferences of the NCPP, the participants educated one another about public policy issues and the various viewpoints of strip-mining and air-pollution regulation. This process exemplifies one view of democratic citizenship, one in which, ideally, citizens learn from one another through discussion and participation in deciding public issues. Of further interest is the nature of the interaction among the numerous interest groups in America. The NCPP example shows that not only do interest groups influence government but government also influences the behavior of interest groups. One can then examine the possibility that in economically stressful times, the incidence of cooperation among interest groups and agencies of government might increase.

I am thankful for the financial assistance I received to conduct this research project. The Government and Law division of the Ford Foundation provided me with a grant to begin the project, and support was later continued by a grant from the Russell Sage Foundation. I am grateful to Marshall Robinson, James Pendleton, and Allan Pulsipher for their interest in this work, and for the fact that the Humanities Institute of the University of Illinois at Chicago provided funds for a leave from teaching so that I could begin drafting this book. The Institute for Government and Public Administration of the University of Illinois supported putting the manuscript on a word processor, and the institute's secretary Shirley Burnette did an excellent job, as did my other word processors, Laurie Kalmanson and Donald Nagolski.

I also wish to acknowledge the assistance of the Resources for the Future Foundation, which provided me with office assistance to begin the project, and assistance from the Government Studies Program of the Brookings Institution, which later provided me with an office to help me continue my research in Washington, D.C. My aunt and godmother, Miss

Flora Roth Symons, greatly assisted me by kindly providing free lodging whenever I returned to Washington, D.C., to conduct follow-up research.

Paul Quirk encouraged me to embark on this project and gave me very useful criticism and advice. Jeffrey Berry and Christopher Bosso also read the manuscript and gave me useful criticism as well as support. Barbara Ferman and Paul Brace gave me bibliographical advice. Paul E. Peterson and the late J. David Greenstone helped in maintaining my career in political science, and my student Barbara Maria Yarnold helped inspire me to work further. I am honored to serve in the same profession with these and other outstanding political scientists.

1

An Experiment in Interest Group Theory

After reading about lengthy and bitter conflicts between businesses and citizens' groups over the environment, many people no doubt wonder whether it is possible to bring opposing groups together to discuss an issue rationally. But if the adversaries met outside the courtroom, if they avoided bruising legislative struggles, could they not reach compromises? Wouldn't this possibility result in more timely economic decisions and thereby save the public a good deal of money? Don't the Japanese follow such a procedure whenever possible and give their economy a competitive advantage over the United States?

The 1977–1978 National Coal Policy Project (NCPP) is the best case to examine for answers to such questions. The NCPP was an elaborate attempt to mediate political conflicts between the coal industry and environmentalists through conferences that involved both corporate executives and environmental lobbyists. A group of individuals founded the NCPP as an experiment, and they spent $1.4 million trying to make it work.[1] Business donated about 60 percent of the money, and the remainder came from the federal government and several foundations.[2] In addition, leaders of the project estimated that participants put in 15,000 days of work, half of which were paid for by businesses in the form of salaries of executives on released time.[3]

At first, this exercise in group cooperation looked like a big success—the business people and environmentalists reached so many agreements, 800 pages were needed to describe and explain them all.[4] Representatives of groups, acting independently of established political institutions and government, did manage to resolve difficult policy issues. But to be truly effective, the agreements required legislative action to enact them into law. Although the executive branch did enact a small part of the NCPP's program in the form of regulations, the government ignored about 90 percent of the agreements. From this perspective, the project must be judged a failure; in the long run, however, it may encourage more sophisticated efforts in group cooperation.

The NCPP should be considered an experiment, an institution designed by its founders to find new ways to reconcile conflict among contending groups. Political scientists rarely have the opportunity to conduct such experiments among participants in politics, and when time and money are spent in an attempt to achieve institutional innovation, it is important for those scientists to describe and analyze the effort. The NCPP deserves to be remembered and understood by lawyers, political scientists, and other scholars of American institutions as well as by business executives, regulators, and environmentalists concerned with environmental and energy policy-making. This book is an effort to make sure that the NCPP is not lost to the general, public record.

PERSPECTIVES ON THE NCPP

A number of different perspectives on the NCPP developed. In 1978 the project received national recognition as a new way to solve disputes and to promote economic efficiency when the report of the NCPP was page-one news in the *New York Times* and the subject of a laudatory article in *Fortune* magazine.[5] In 1979 prestigious general reports on America's energy future issued by a group of Harvard scholars and by a leading Washington research institution, Resources for the Future, both praised the NCPP as a path-breaking effort to design more effective institutions in energy policy-making.[6]

But by the middle of 1980 the NCPP was often considered a failure, as people active in the environmental and energy policy networks realized that few of the project's recommendations were to be implemented.[7] Indeed, most of the participants in the NCPP itself were disappointed by the relatively few cases of government adoption of their recommendations.

Still, a more positive view of the NCPP began to appear a short time later. People interested in further experimentation with new institutions for dispute resolution sometimes cited the NCPP as a pioneering effort that, although having only a modest impact on policy, encouraged others to develop more focused discussions and negotiations to adopt or reform policy regulations.[8] Interest in regulatory negotiation, policy dialogues, environmental mediation, and other forms of "alternative dispute resolution" continued during the 1980s, and some scholars now view the NCPP as an early experiment in a continuing line of institutional experimentation.[9]

There were a number of other evaluations of the NCPP. Some environmentalists were critical of the NCPP, viewing it as an institutional means to "co-opt" the environmental movement by making it less aggressive in its demands and by persuading it to give up some of its hard-won gains in Congress.[10] Sophisticated versions of this criticism maintained that litigation is a particularly efficient means to use the scarce resources of the environmentalist movement. In litigation both sides have an equality of standing before the judge, even in those cases when environmentalists have less money to pursue legal research.[11] On the other hand, some of the leaders of the coal industry in 1980 seemed to have held an analogous view of the NCPP. Some members of the National Coal Association thought that the NCPP compromised the bargaining position of industry and would trade away the right to lobby for changes in strip-mining regulations to favor the industry.[12]

Yet another evaluation of the NCPP in the early 1980s held that it was a useless enterprise, "a waste of time." Some environmentalists thought that the output of the NCPP was not worth the time put into the enterprise by environmental lobbyists, whose expertise was in great demand.[13] The attitude of others in the energy and environmental policy networks could be summarized by a comment by John O'Leary, an official in the Department of Energy under President Carter, who said, "I never heard of it [the NCPP], and if I had, I wouldn't have paid attention to anything coming out of that sort of thing."[14] Coal policy-making has been exceptional for the bitterness of policy disputes, and some people considered it naive to engage in policy conferences about issues that frequently invoke bitterness on all sides.[15]

Was the NCPP a promising experiment in conflict resolution, an abject failure in policy-making, a trap for unwary negotiators on both sides, or simply a foolish waste of time and money? Enough years have now passed for some answers to emerge, and to develop these answers, one needs to work with a few closely related ideas called "cooperative pluralism."

COOPERATIVE PLURALISM

"Pluralism" has come to mean the view that the group life of a society is very significant, especially in terms of the relative significance of groups and the state. Accordingly, the term readily applies to the NCPP, an experiment in which the importance of groups was affirmed in relation to the government.

In the type of pluralism exhibited by the NCPP, the power of producer groups, such as coal companies and electric utilities, is partially checked by countervailing power groups representing interests other than those of the dominant producer groups in a given policy area. In this case, the countervailing power groups were mostly environmentalist lobbies. But it is not only so-called public interest groups that exercise countervailing power in public policy. Labor may check business, or the power of a well-organized producer lobby—say those favoring continued production of high-sulfur coal in West Virginia, Ohio, and Illinois—may be partially checked by another producer lobby—say those favoring higher production of low-sulfur coal in Wyoming and Montana.[16]

The NCPP included both producer groups and countervailing power groups, but purposely excluded representatives of government. This exclusion must be understood in the context of changing views of the relationship of interest groups to government in the last forty years among political scientists who might be called "pluralists."

In their strong reaction against legal philosophies emphasizing the sovereignty of the state and against a political science based on the textual analysis of legal documents, the first pluralists tended to argue that the lobbying power of interest groups tended to dominate the actions of the state. Political leaders were seen as umpires or referees who organized and channeled political struggles whose outcome was determined by the power balance among the group players in the game of politics. This view was most widely held in the 1950s, and found such adherents as Arthur F. Bentley, Earl Latham, and, in general, David B. Truman.[17] One can say that the leaders of the NCPP were gambling that the Bentley-Truman theory of groups was true, for the theory predicted that a coalition of business groups and environmentalists would suffice to determine public policy.

In the 1960s, another group of scholars, based at Yale and including Robert A. Dahl, Charles E. Lindblom, Nelson W. Polsby, and Aaron B. Wildavsky, were also called pluralists, but they perceived American politics differently from the Bentley-Truman view.[18] The Yale pluralists saw American politics as being very decentralized, not substantially controlled by the state or by a powerful elite. Within this decentralization, the Yale pluralists saw interest groups as having an influential role, but they did not argue that groups somehow controlled government. In fact, the Yale pluralists stressed that government leaders themselves had influence and might influence interest groups, as well as being influenced by them.[19] The leaders of the NCPP generally agreed with this outlook, but they wanted to reduce the influence of government in the political process by forming a coalition of groups.

In the 1970s, another pluralist school appeared. It affirmed the significance of groups in the political process but differed from its predecessors by portraying group influence as having serious negative consequences for democracy.[20] Such writers as Theodore Lowi and Mancur Olson saw state policy fragmented into hundreds of separate arenas, each policy controlled by coalitions dominated by producer groups and usually including the politicians and government executives acting in the area.[21] In American politics, this meant control by "subgovernments," cooperative coalitions of producer groups, congressional committee members, and government agencies acting together to promote the economic interests of the producer group. For instance, tobacco growers worked with key congressional committee members and officials of the Department of Agriculture to maintain federal subsidies for tobacco production.[22] This view might be termed "multiple-elite pluralism," because the importance of groups lies in their providing the organizational basis for a multiplicity of separate elite subgovernments, each controlling its own policy sector.[23]

The Bentley-Truman pluralism and the Dahl-Lindblom pluralism assumed the importance of countervailing power groups. But multiple-elite pluralism argues that countervailing power groups are generally absent. The participants in the NCPP disagreed with this perspective; virtually all of them considered environmental lobbies to be influential. Consequently, cooperative pluralism should not be confused with multiple-elite pluralism. Cooperative pluralism presupposes the existence of countervailing power groups.

I adopt a view of pluralism developed by political scientists in the 1980s that will help us to understand both the NCPP and the idea of cooperative pluralism. Recent pluralists find that producer groups, countervailing power groups, and agents of government each normally influence the policy-making process in different areas.[24] Such writers would be surprised if producer groups and countervailing power groups were the sole agents of influence, without governmental agents having some separate influence. Although this unusual situation was the goal of the NCPP, it did not show how a coalition of producer groups and countervailing power groups could eliminate the separate influence of government agencies within a policy area, and indeed its effort to exclude government influence failed. On the other hand, this failure confirmed the new pluralist expectation that governmental agencies autonomously influence the policy process and are not simply influenced by political groups.

Research now finds producer groups, countervailing power groups, and autonomously acting government normally present in particular American policy-making processes. Therefore, cooperative pluralism does *not* refer to the producer/government coalitions of the multiple-elite theory, now seen as atypical. Instead, cooperative pluralism refers to cooperation among producer groups, countervailing power groups, and autonomously acting government, the three typical actors within a particular area of public policy. If coal companies, environmentalist lobbyists, and the Office of Surface Mining interact, for instance, the term "cooperative pluralism" refers to the cooperative aspects of this interaction.

The relationships among groups and government officials within an issue area are characterized normally by significant conflicts of interests, and the resulting politics forms part of the domain of political scientists. But I argue here that cooperative aspects of such relationships exist or can be developed.

AN EXPERIMENT IN COOPERATIVE PLURALISM

The outstanding feature of the NCPP was experimentation in cooperative pluralism. While participants were neither political scientists nor theoreticians, the NCPP was well designed to enlighten us about pluralism. Its leaders took for granted the importance of the interaction of coal companies and other economic producers with the countervailing power organizations of environmentalist lobbies. A major question of the participants concerned the role of government, which they hoped to resolve by excluding government from their negotiations altogether. They thought it would be easier to reach decisions if government officials were kept out; they believed a united front of producer groups and countervailing power groups could control the enactment and amendment of federal regulations. It turns out, however, to be very difficult to gain support from group leaders outside of the negotiating conference, and therefore the participation of government officials is very useful to stabilize and to implement negotiations among group representatives. In this way, the NCPP "experiment" eventually demonstrated the importance of state participation in bargaining processes.

In the history of the NCPP, this empirical observation is related to a normative point of democratic theory. The NCPP was formed so that the opposing sides of business and environmentalists could discover common interests. The NCPP experiment in cooperative pluralism indicates that appropriate institutions can help interest group leaders to find common goals and to learn other aspects of public policy. The participants taught one another to produce an elaborate platform that stated newly learned common interests, and some of them felt that this experience was the most valuable part of the NCPP.

In *Considerations on Representative Government*, John Stuart Mill emphasizes the need for political institutions to develop the intellectual and moral faculties of political participants as well as their capacity to make and implement public policy efficiently.[25]

> We have now, therefore, obtained a foundation for a two-fold division of the merit which any set of political institutions can possess. It consists partly of the degree in which they promote the general mental advancement of the community, including under that phrase advancement in intellect, in virtue, and in practical activity and efficiency and partly of the degree of perfection with which they organize the moral, intellectual, and active worth already existing, so as to operate with the greatest effect on public affairs. A government is to be judged by its action upon men, and by its action upon things;

> by what it makes of the citizens, and what it does with them; its tendency to improve or deteriorate the people themselves, and the goodness or badness of the work it performs for them, and by means of them.[26]

Mill would have lauded the NCPP as an institution promoting "the general mental advancement of the community," as the participants learned a great deal not only about coal policy, but about their own common interests, seen against the background of conflicting interests.

Mill underscores the value of individuals and representatives learning to transcend the strife of individual competition through political participation. If the individual does not learn in "this school of public spirit,"[27] then such a "man never thinks of any collective interest, of any objects to be pursued jointly with others, but only in competition with them, and in some measure at their expense."[28] On the other hand, Mill affirms

> the moral part of the instruction afforded by the participation of the private citizen, if even rarely, in public functions. He is called upon, while so engaged, to weigh interests not his own; to be guided, in case of conflicting claims, by another rule than his private partialities; to apply, at every turn, principles and maxims which have for their reason of existence the common good; and he usually finds associated with him in the same work minds more familiarized than his own with these ideas and operations, whose study it will be to supply reason to his understanding and stimulation to his feeling for the general interest.[29]

A second outstanding aspect of the NCPP is that it embodied Mill's theory of democratic participation. As such, it performed functions similar to those performed by American legislatures, which are seen as educating participants about public policy and common interests, as described by Mill's contemporary student, William K. Muir.[30] He described the California legislature as a school in which the legislators taught one another about various social interests and their relationship to public policy.

Writers who are not influenced by liberal individualism and who are not especially influenced by John Stuart Mill and Robert Dahl may nevertheless come to conclusions similar to those that I label "cooperative pluralism." An outstanding example of this is Jane Mansbridge's work *Beyond Adversary Democracy*.[31] Influenced by Aristotle and Rousseau, Mansbridge emphasizes citizenship, participation, and equality as they emerge in face-to-face interactions among citizens who are, unlike the legislators and interest-group leaders studied by the cooperative pluralists, not from the political elites. Mansbridge does, however, emphasize the importance of mixed relations of conflicting, "adversary," and common, "unitary," interests within political interactions. She argues that modern American politicians and political scientists are insufficiently aware of the extent to which common interests can affect political interaction. She calls for political institutions to develop further the common interest dimension of politics, though she does not deny that politics often involves conflicting interests, served by present institutions designed to formulate legitimate policies from conflicting interests. Like Muir, Mansbridge believes that politics sometimes can involve discussion, learning, and an exchange of ideas among the participants. It is striking that the partisans of Rousseau and Mill can agree on the importance of these ideas and the need for their appreciation by other Americans.

The third outstanding characteristic of the NCPP is its demonstration of an institutional mode to discover common interests to enhance economic efficiency while respecting public values such as environmental preservation. During the past few decades, of course, Americans have been increasingly concerned to find ways to compete effectively in international trade. Indeed, in 1977 this concern was a major justification by the leaders of the NCPP for their experiment in intergroup negotiation. The NCPP showed a talent for such negotiation, for many agreements were reached. Such agreements could increase economic efficiency by cutting costs imposed by long delays and other effects of political controversies between business and its opponents, such as environmentalists or property-holders opposed to nearby construction. On the other hand, subsequent events demonstrated that the direct intergroup negotiation practices of the NCPP were insufficient to gain stable agreements and that state officials must be added to attain such stability.

The NCPP pointed the way to cooperative pluralist negotiations—among producer groups, countervailing power groups, and autonomous government—to promote economic efficiency and to enhance other public values. Such negotiations may offer an important opportunity for American society, but cautious appraisal remains necessary. Another twenty years of experimentation may be needed to gain experience with cooperative pluralist negotiation techniques. Even then, institutions for cooperative pluralism may provide only a small part of the needed institutional changes for renewed economic competitiveness. Nevertheless, an increase in cooperative negotiation would be an important gain for American society, and support for the development and appraisal of such institutions is needed.[32]

The NCPP thus was an experiment in three phases of cooperative pluralism, and should help us learn more about the concept. Of course, other political scientists have published case studies and theoretical observations that bear upon cooperative pluralism. James Q. Wilson and his students completed several studies that show countervailing power offsetting producer interest groups and providing a possibility for agency autonomy, as professional values of government officials become an independent variable affecting policy.[33] William K. Muir has studied the California state legislature as an institution whereby the legislators learn and teach one another about public policy.[34] Finally, a literature is emerging on coalitions in American cities between local government and interest groups seeking to develop policies that enhance economic growth.[35]

To avoid confusion, we must distinguish cooperative pluralism from "corporatism," or corporatist decision making, a subject that has recently gained much attention from political scientists.[36] Democratic corporatism refers to decision-making processes in a number of European countries—Germany, Austria, Switzerland, the Netherlands, Belgium, Sweden, Norway, Denmark, and Finland. Interest groups in corporatist countries form national federations, which are centrally governed, and which are "exclusive" in the areas of representing business or labor. Corporatist decision making occurs when the single national business federation and the single national labor federation meet together with representatives of the national government to set guidelines for a nation's policy on macroeconomic policies, such as new taxes, the rate of wage increases, selection of industrial sectors encouraged to export in competition with other nations, and so forth.[37]

Cooperative pluralism refers to cooperation on a more limited scope than corporatism, within a single policy area, such as the regulation of stripmining, rather than broad-scale cooperation over a society's macroeconomic decisions. The two concepts do overlap, however. Some writers have described a middle-level corporatism in some European countries, in which regional economic planning is conducted through corporatist negotiations.[38] Several political scientists are currently investigating whether American city and state economic development decision making sometimes exhibits a type of local corporatism.[39] Regularly structured regional negotiations among business, labor, and government over economic development might be characterized either as cooperative pluralism or as middle-level corporatism. The history and experience of the NCPP, and the lessons that can be drawn from it, are useful for understanding the theory of cooperative pluralism.

2

Coal, Strip-mining, and Air Pollution: The Policy Context

An understanding of the technical and political aspects connected with the complex issues of coal, strip-mining, and air pollution is necessary for any discussion of the NCPP experiment in cooperative pluralism. The early 1970s, immediately before the formation of the National Coal Policy Project in late 1976, was a time of political transformation for a large range of related energy and environmental issues. The first generation of protest in the 1960s focused on civil rights and the war in Vietnam, but by 1969 a second generation of protest brought up new issues, including women's rights, consumer protection, government reform, and environmentalism.[1] National attention to the Santa Barbara oil spill foreshadowed a sudden rush of environmental concern, both by the general public and by governmental officials who believed that action on environmental issues would be a relatively cheap way to please a large sector of the public. Established conservation groups, like the Sierra Club, rapidly expanded their membership and became more aggressive "environmental" lobbies.[2] In 1969 and 1970, important new environmental groups appeared on the national scene, often financed by grants from foundations given to young, socially conscious lawyers, who saw an opportunity to affect public policy through public interest law suits.[3] The media, especially television news, rushed to report environmental issues and often provided good pictures of nature being despoiled, dramatic stories of conflict between citizens and polluters, or shocking tales of deathly risks from chemicals or seemingly innocuous products. Such news stories bridged the concerns of a varied audience, which was otherwise viewing divisive and resentment-producing descriptions of radical protest, war in Vietnam, new social movements, and the bitter backlash against such movements.

Politicians, like media executives, saw the potential of appealing to an otherwise divided public through the passage of environmental measures. In 1969 the federal government passed the National Environmental Policy Act, which mandated environmental protection statements for federal construction projects.[4] This act also established the President's Council on Environmental Quality, and in 1970 the Environmental Protection Agency (EPA) was established. The Clean Air Act amendments of 1970 were a major extension of regulations, mandating new controls on the burning of coal and other industrial emissions.[5]

The environmental movement was also very active at the local level: indeed its local impact has been perhaps as important as Washington lobbying and legislation.[6] In the early 1970s thousands of local environmental protection groups appeared, most of them short-lived, and many of them directed against the construction of new public works, especially the highly unpopular nuclear electric power plants. In addition, environmentalists pressed for restrictions on economic development and various forms of construction and argued for protecting nature in the form of greenbelts, parks, and wildlife protection.

The environmental movement acquired power from its writers and activists, national and state organizations, lobbies, favorable media coverage, friendly politicians looking for popular political initiatives, major foundation support, active litigation, new legislation at all levels of government, and initial appeal to many social groups, even Republicans and conservatives. In 1969–1970, the environmental message spread at top speed, carried everywhere by sympathetic journalists, friendly groups, and politicians seeking support. The onrush of environmentalism was so sudden that the businesses that were to be regulated were caught by surprise and were unable to define their objectives (as individuals, they cared about the environment, too) or to mobilize a lobbying campaign in support of their objectives quickly enough to block the passage of the new national and local environmental laws and regulation.[7]

By 1971 a whole new dimension of American politics had appeared in the form of the environmental issue. This dimension was itself complicated in the fall of 1973 by the sudden impact of "the energy crisis," with the formation of the Organization of Petroleum Exporting Countries (OPEC), the Arab-Israeli war, the export boycott in petroleum, and the appearance of gas lines.[8] In 1974 political leaders were concerned about "energy independence" from oil imports from OPEC, and particularly from the producers of the Middle East. Suddenly, not only was the country to try more strenuously to protect the environment, but simultaneously to push hard for independent production of energy. To the environmentalist, this meant a crash program in national energy conservation—serious regulation of auto construction for gasoline economy; constant monitoring of the use of electricity and other forms of energy in industrial production, commercial buildings, and home heating; and development of new, nonpolluting alternative forms of energy, such as solar sources, windmills, and biomass, among others. Environmentalists insisted that the construction of nuclear power plants for electricity be stopped, and that plants in operation be closed, even though nuclear energy provided about 12 percent of the nation's electricity by the late 1970s.[9]

With the convergence of the energy and environment issues, the policy area of mining and use of coal assumed greater importance to policymakers and opinion leaders. In 1967–69 coal had received priority on the agenda of issues during controversies about mine safety regulation.[10] After the oil boycott and the OPEC price increase, pundits were fond of noting that "America is the Saudi Arabia of coal." Could not American coal production be increased to stem the imports of oil? But such an increase would require the development of new energy technology. Petroleum is mainly used in liquid form for transportation, but the transformation of coal into liquid fuel would require massive development projects.[11] Furthermore, although using coal to generate electricity or to heat industrial plants could save some oil, burning coal is much more polluting than burning oil or natural gas. As coal mining had been moving in the direction of strip-mining, such proposals implied a great increase in such mining practices, particularly in the West, a project which provoked fierce opposition from environmentalists.

The addition of the energy crisis of 1973 to the newly perceived environmental problem further perplexed and irritated many business leaders. OPEC had demonstrated a troubling capacity to increase greatly the cost of oil. Meanwhile, it seemed difficult to find substitutes for petroleum. Without a great deal of publicity, business leaders gave great priority to finding means to conserve energy, particularly electricity, but also petroleum.[12] Industrial conservation achieved great success by 1980, but this success was not apparent in the middle 1970s. Some business leaders were still enthusiastic about the possibilities of producing electricity with nuclear power plants, but such efforts were meeting increasingly vociferous and well-organized opposition from environmental groups and local citizens' groups. Domestic natural gas appeared to be an effective substitute for oil in industrial steam processing and in generating electricity, but its price was still regulated by the federal government, conversion from oil to gas would be expensive in some cases, and many energy experts felt that burning gas to heat water for turbine steam was an inefficient use of this nonpolluting substance.

Virtually all experts agreed that American oil production was bound to decline slowly, because the country had been so thoroughly prospected for oil.[13] Development interests pressed for less environmental restriction of offshore and Alaskan oil production, but such development would only slow the rate of decline. The development of coal liquefaction and coal gasification might offset some of the decline, but would be very expensive and therefore engender great political opposition. Accordingly, business leaders concerned about energy could not delineate a general course of action and felt that no matter what policy they proposed, a powerful coalition of environmentalists and officeholders would oppose energy development programs.[14]

Business leaders were correct. Environmental lobbies, various litigation groups, and elected officeholders were ready to do just that. Although liberal politicians were ordinarily more enthusiastic about restricting energy development projects, local moderates and conservatives in public office might support organizations of local property holders opposing offshore drilling, the construction of nuclear power plants, and coal-burning utilities. In the middle 1970s, environmental lobbies and litigating groups had become relatively powerful. Although, as David Vogel observes, business was then beginning to mobilize its forces against the environmentalists and would achieve more success in the early 1980s, in the middle 1970s, the full panoply of environmental lobbying, litigation, and electoral participation was unleashed, and it appeared that environmental considerations would receive serious political backing in most major energy development cases.[15] By 1976 it was clear that almost all development of nuclear power beyond projects in construction was likely to be blocked by environmental coalitions. In any event, even if developers eventually received necessary government permits, they would often be delayed for years by litigation or national-state-local legislative roadblocks, while at the same time the cost of construction increased greatly due to inflation.

Government agencies increased their autonomy and influence during this period. The EPA was not simply a conduit for business or for environmental pressure groups, but was able to advance an independent position on some issues.[16] Under President Ford, the Energy Research and Development Agency, and its successor agency under President Carter, the Department of Energy, began to initiate research independent of pressure groups in the area of alternative, experimental sources of energy such as new coal technology, fusion, and solar sources.[17] The Atomic Energy Commission, generally regarded as a promoter of nuclear energy in coalition with interested corporations, was broken up and its regulatory functions transferred to the Nuclear Regulatory Commission, generally friendly to the nuclear industry but not always a dependable supporter of it. The Federal Power Commission, which gave way to the Federal Energy Regulatory Commission under President Carter, often opposed the requests of industry in setting prices of natural gas and in regulating interstate transmission of electricity. In short, the autonomous influence of government agencies upon energy and environmental policy increased during the years 1969–1976 before the organization of NCPP.

The new range of interconnected environmental and energy issues produced a very complex policy environment and created a great deal of confusion and frustration among interested parties and the general public. This confusion was no doubt inevitable given the rapid flux of changing events—the surge of environmentalism, the energy crisis, the organization of new environmental pressure groups, and the expanding power of federal agencies. Most situations involving energy and environmental policy did not exhibit the second-stage pluralist phenomenon of a controlling triangle of interest group, legislative committee, and government agency. Instead, the prevalent pattern revealed a triadic power alignment of producer groups, countervailing environmental groups, and an autonomous government agency.[18]

Unsurprisingly, some observers proposed that elements of cooperative pluralism be added to the complex, confused energy and environmental situation. By the middle 1970s, it was apparent that all three power groups would have a permanent influence on energy and environmental policy. Furthermore, the energy and environmental policy areas were fraught with national security considerations after the 1972 energy crisis. Clearly, there was a need to resolve some of the confusion and conflict in these policy areas. Although most policy-makers considered it extremely idealistic, the argument for greater cooperative pluralism in energy and environmental policy was much stronger by 1977 than it had been ten years earlier. The foundation had been laid for the NCPP.

COAL

Coal had been the country's main source of energy in the early twentieth century, but its importance dwindled after World War II with the advent of the gasoline-powered auto and the switch in industrial, commercial, and home heating from coal to oil or natural gas. By the 1970s coal was used primarily to produce steam to generate electricity, and at that time, about half of the nation's electricity was still derived from processes based on burning coal. Conversely, about 70 percent of the coal produced in the United States was used to generate electricity, with most of the rest going to industrial steam processing for heating, for operating machinery, and for producing steel.[19]

In the middle of the century American political leaders and public opinion leaders did not often consider coal policy. While coal-related issues might be important in producing states such as West Virginia and Kentucky, coal made the news mainly because of sporadic wildcat strikes or disastrous mine accidents. In fact, the 1950s were a time of relative labor peace, as John L. Lewis, leader of the United Mine Workers (UMW), negotiated an agreement with leading coal companies conceding that labor would not block the move to strip-mining and other techniques that greatly reduced the number of mining jobs. In return, the coal companies established a relatively encompassing pension fund that would pay benefits to the many miners who lost their jobs.

By the middle 1960s, however, criticism of the mediocre safety record of the coal industry became widespread. After a hard-fought struggle, Congress passed the Coal Mine Health and Safety Act in 1969 to establish new standards for control of explosive coal dust and mine gases and to provide for safer mining practices. This law also established a system of modest pensions for miners disabled by the dreaded "black lung" disease.[20] Since the law was passed, mine safety has not received much attention from national decision makers (although one heard occasional protests that the Reagan Administration inadequately enforced the safety standards).

Of course, strip-mining without controls scars the land with large pits and heaps of removed dirt. More disturbing, farmland or wildlife habitat may be destroyed, upstream dirt piles may cause flooding from runoff or themselves become the source of dangerous mudslides, water running through the dirt "spoils" may become acidified through leaching minerals and thus may kill aquatic life in streams, and the underground water table may be disrupted creating problems for rural residents dependent on well water. Uncontrolled strip-mining is an environmental disaster. I was shocked to see the scars on the land wrought by strip-mines visible from a plane flying over Appalachia in 1970. It is no surprise that control of strip-mining was one of the biggest environmental issues of the 1970s.[21]

Following the energy crisis of 1972–1973, and in the years immediately preceding the formation of the NCPP, the attention of national public leaders turned to coal. Some cited the enormous quantity of coal reserves in the United States and averred that this country should develop these reserves to gain energy independence from oil imports. Upon examination, however, it was not clear what specific policies would work. Some power plants using oil could have their boilers converted to coal burning, but in the middle 1970s only about 12 percent of the nation's electricity came from oil-fired plants. Some industrial boilers might be similarly converted, but such conversions are often expensive. Further, burning coal produces more pollution than burning oil, and new clean air laws would require the installation of expensive antipollution equipment.

Enthusiastic proponents of increasing coal production called for an increase in strip-mining in the Great Plains fields of Wyoming, Montana, and North Dakota. The additional coal might be used for massive coal gasification projects, in which coal or peat is heated until it changes to gas. Such gas could be piped for heating, industrial use, or electricity generation. But how much petroleum would be saved? And such coal gasification projects would have major negative impacts on the environment, especially through the expansion of strip-mining.

Opinion leaders realized that coal development would have its most important impact in meeting the demand for additional electricity, which could be generated from coal burning rather than from petroleum or nuclear sources. Judging from opinions expressed by NCPP participants from the chemical and aluminum industries, leaders of industries intensively using electricity were concerned in 1976 that if nuclear- and petroleum-based power were not developed, then coal production must be increased to prevent brownouts in the early 1990s. Environmentalists, on the other hand, argued that instead of building new power plants, conservation should be emphasized.[22] In the short run, the environmentalists proved correct; brownouts did not occur due to effective conservation by industry following the incentives of higher electricity prices. Still, one wonders whether there will be shortages of electricity at the turn of the century, as some areas use up surplus generating capacity.

At the time of the formation of NCPP, then, confusion reigned over the direction of national coal policy. Politicians and opinion leaders had to master the web of connections among the policies governing coal production and other energy sources, environmental problems, imports and national security questions, and labor issues. These tangled relations produced confusion among politicians and conflict among interest groups.[23]

STRIP-MINING

A great part of the work of the NCPP concerned the strip-mining of coal, an issue that emerged on the national congressional agenda first in 1968 and again in 1971, and that was fought over for several years, leading to the passage of the Surface Mining Control and Reclamation Act of 1977 (SMCRA). The first significant use of strip-mining technology to mine coal occurred in the middle 1950s. Its use spread rapidly; by 1971, half of America's coal was strip-mined, and the proportion has continued to increase since then. Somewhat more accurately termed "surface mining," as opposed to deep mining or underground mining, strip-mining uses large earth-moving machinery, such as huge shovels, bulldozers, and giant trucks, to remove the earth covering coal seams relatively close to the surface. Surface mining may consist of digging pits in relatively flat ground, or it may be used to cut off the sides of hills or even to cut off the tops of hills. Most surface mining is coal mining, but it is also used to mine other minerals, particularly copper.[24]

Obviously, strip-mining has brutal effects on the environment. The national concern for the environment after 1969 greatly energized the movement to control strip-mining, the mine technology which was displacing the traditional deep-mining underground techniques.

Political scientist Walter Rosenbaum noted in 1985 that a million acres of strip-mined land "remain a wrecked and ravaged waste, long abandoned by its creators."[25] In the normal course of American politics, surface mining would never be permitted in an area inhabited by the politically organized middle class, such as Long Island, New York, the state of Massachusetts, or greater Seattle, for instance. However, coal is located in poverty-stricken areas of Appalachia, where there is no stable source of jobs,[26] in relatively thinly settled rangelands in southern Montana and northern Wyoming, and among the family farms of southern Illinois and western Kentucky. The political weakness of opponents of strip-mining in these areas, combined with the attractions of economic development to areas without much wealth, made it politically possible to conduct surface mining of major coal seams.

In keeping with the American tradition, state government initially regulated strip-mining.[27] Basic regulatory proposals might occur to anyone: if surface mining is not outlawed, then require the miner to restore the land to its natural contour, rather than leaving huge pits and scars. Another obvious regulation would prohibit or closely regulate strip-mining in areas where it might disrupt the water table, cause flooding of mountain streams, or destroy valuable farmland, for example. Another reasonable proposal might suggest a tax on current strip-miners to pay for the reclamation costs of abandoned strip mines. But in the 1960s the strip-mining control policies of state governments usually confirmed the pessimistic theories of the multiple-elite pluralists, though laws were passed in almost all involved states to control surface mining. But these laws were what Edelman called "symbolic" legislation; the laws existed as a symbol that government was acting, though in fact, state strip-mining control was unevenly enforced, for the political pressure of strip miners on government remained steady, while the political organization of strip-mining reformers would wax and wane. As multiple-elite pluralists E. E. Schattschneider and Grant McConnell emphasized, one needed to expand the forum of decision making to the national level to defeat the locally powerful strip-mining interests.[28]

First, local coalitions of citizens, conservationists, and politicians organized to limit strip-mining, such as the group Save Our Kentucky in the early 1960s. Soon national conservationist groups such as the Sierra Club, the Wilderness Society, the Audubon Society, and the Izaak Walton League opposed strip-mining practices.[29] The first congressional hearings were held on the issue in the Senate in 1968, but no action was taken. Meanwhile, the environmental movement was rapidly spreading, and the new environmental lobbying and litigation groups also opposed strip-mining without regulation. The Coalition against Strip Mining was organized in Washington to lobby for regulation. It consisted of twenty-six national environmental groups and regional anti-strip-mining federations. The coalition held that surface mining should be entirely outlawed.[30]

In the late 1960s, it had become clear that strip-mining was likely to become widespread in the Wyoming-Montana fields and in a few other western locations. This activated a new constituency opposed to surface mining, including prosperous ranchers concerned about the future of the land, western conservationists and outdoors hobbyists, and local politicians.

With the surge of environmentalism in 1970 and with the burgeoning strength of the coalition opposed to surface mining, the chances for passing a national law controlling strip-mining appeared to be high. The Nixon administration decided to try to capture the issue by proposing relatively modest legislation. In his State of the Union address on February 15, 1973, President Nixon stated: "New legislation with stringent performance standards is required to regulate abuses of surface and underground mining in a manner compatible with the environment."[31] In 1972 the National Coal Association recognized political reality and decided not to oppose strip-mining control legislation outright but to lobby for weakening amendments.[32]

This is not to say that the leaders of the coal industry were an accommodating lot. Quite the contrary. They were a particularly hard-bitten group. The coal industry had a history of particularly angry and bitter labor conflict, although this had been mitigated in the 1950s and 1960s. (Still, there was a long strike in 1977.)[33] Corporate mine owners did not live near their mines, and thus community ties did not deter them from wrecking the environment.[34] Moreover, coal was a competitive industry in which owners were very motivated to cut costs by whatever means possible, especially by expanding surface mining.[35]

The scene was set for intense political conflict. Strip-mining control could be very expensive for the miners. For instance, a widely supported idea to eliminate strip-mining on hillsides having greater than 20-degree slopes would have eliminated the mining of 51 percent of surface-mined Appalachian coal, according to the President's Council on Environmental Quality.[36] On the other hand, environmentalists had especially intense views in opposition to strip-mining; indeed, their goal was to eliminate it entirely. A bill to this effect got sixty-nine votes in the House in 1974.[37]

THE PASSAGE OF STRIP-MINING REGULATION

By 1971 the passage of a federal surface-mining control act seemed highly probable. Environmentalism had become a highly popular political position, strip-mining control was high on the list of environmentalist priorities, and all significant environmentalist lobbies supported such legislation. To be sure, there were major differences in Congress about how strict strip-mining legislation should be, or, put another way, how much such regulation should cost the coal industry. Five positions could be discerned in subsequent debate and voting, which I list here in order of increasing strictness. First, as a matter of strategy the National Coal Association admitted that a law was needed, but in fact coal lobbyists wanted such a law to be as weak as possible given the political circumstances. Second, though the Nixon and Ford administrations wanted some degree of regulation, they differed from the congressional majority in being willing to permit strip-mining on hilltops, near water tables, and on farmlands, and in wishing to exempt a greater number of small mines.[38] The third position, eventually written into law as SMCRA in 1977, might be described as "moderately strict," going further than the Ford administration on issues such as the above.[39] A fourth position, identified with some congressional liberal Democrats, would impose quite substantial costs on the coal industry by precluding strip-mining on large blocks of land. For instance, 136 members of the House and 28 Senators voted in 1975 to forbid strip-mining on hill slopes of greater than 20 degrees, thereby putting perhaps half of all Appalachian coal off-limits.[40] The fifth position, supported by virtually all environmentalists and by 69 members of Congress voting in 1974, would have made surface mining illegal altogether.[41]

A middle position, SMCRA-type bill first passed the House in October 1972, right before adjournment, by a vote of 265–75. In 1973, still lacking a majority, the opposition to strip-mining control used stalling tactics and multiple amendments to try to delay or kill a moderately strict bill. Republicans on the mining subcommittee of the House Interior Committee repeatedly refused to appear for votes on the legislation in subcommittee, which then had to adjourn for lack of a quorum.[42] When the legislation finally reached the floor of the House in July 1974, six days of debate were held to consider various amendments, the longest debate on the House floor on any measure for four years. However, a majority turned away a weaker substitute bill by 156–255, and the overall measure finally passed 291–81. The Senate had previously passed a similar measure by 82–8.[43] The conference to resolve the differences in the bills proved especially difficult; it met twenty times and did not report until December 3, 1974. The conference report passed both houses by voice vote in the next few days.[44]

Although a majority of both houses of Congress favored a moderately strict bill, presidential vetoes delayed passage for three years. President Ford wanted looser, more permissive regulation. Consequently, Ford vetoed the 1974 bill after legislative adjournment so that a veto override vote was impossible (the pocket-veto procedure). In his veto message, Ford stated that the bill would produce "excessive federal expenditures and would clearly have an inflationary impact on the economy."[45]

The aggressive, liberal post-Watergate Congress came to Washington the next month, and in March 1975 passed basically the same legislation by a Senate vote of 84–13 and a House vote of 333–86. The conference bill passed both houses in early May, with a House vote of 293–115. President Ford vetoed the bill for a second time, this time arguing that the measure would eliminate jobs and raise consumer's utility bills. On June 10 the attempt to override Ford's veto in the House failed to get the necessary two-thirds in a vote of 278–143.[46]

The Ford administration and the coal and electric utility industries succeeded in changing the votes of twenty-four Republican members of the House from the conference vote to the override vote. Five Democrats changed their votes to the Ford position, but four changed in the other direction.[47] *The Congressional Quarterly* reports:

> The lobbying campaign [by the coal industry] might not have worked without a parallel effort by the utility companies, not normally thought of as one of the strongest pressure groups in Congress.

> The utilities created a consumer argument against the bill by insisting they would have to raise their rates if it passed. Local utilities contacted their own representatives, and often were able to convert people the coal lobbyists had lost.[48]

In 1976 most Democrats supporting strip-mining control were willing to wait until 1977, for all the major Democratic presidential candidates supported such legislation. In addition, there was a technical wrangle over whether House rules permitted passage of the same bill twice in the same Congress.[49]

Strip-mining legislation was one issue settled in the Ford versus Carter presidential race. Essentially the same bill vetoed by Ford passed in 1977. The Senate voted 57–8 initially, and 85–8 for the conference measure; the House voted 241–64 initially and 325–68 for the conference. President Carter was pleased to sign the measure into law.[50]

The 1977 law remains the basic legislation. After its passage, it was regarded as acceptable by environmentalist lobbyists, though they would have preferred stricter controls. The coal mining industry, not surprisingly, regarded passage as a defeat and immediately sought to block enforcement of the measure through litigation.[51] The law was not well enforced during President Reagan's first term,[52] but there has been a greater effort at enforcement since then with the resurgence of environmentalism.

The six-year lobbying battle before the passage of SMCRA produced bad feeling between coal and utility lobbyists on the one hand and the environmentalists on the other. Accordingly, the NCPP met in 1977 amid an atmosphere of bitterness between the two sides on strip-mining issues. SMCRA was passed while the NCPP was still debating its own strip-mining position, and the coal and utility representatives would not necessarily agree to measures imposed on their industries by the environmental coalition in Congress. The relatively broad scope of agreement in the NCPP on strip-mining issues was a surprise both to the project's participants and to outside observers.

AIR POLLUTION

The first federal Clean Air Acts passed in 1963 and 1966, but at the time the environmental movement took off in 1969, the federal government had taken only tentative steps toward regulation. At that time, Washington only required the state governments to submit plans regarding the control of air pollution, but the federal government had set no standards to measure pollution, nor would it act to enforce the state plans after they were submitted.[53]

The Clean Air Act amendments of 1970 signalled a massive change; this is one time when the federal government did not act in an incremental fashion. The basic standards of the present federal regulation were established in 1970, when this law passed Congress with only weak opposition. This complicated law cannot be discussed fully here, but some aspects relevant to the NCPP will be outlined.[54]

Obviously, burning coal is a major source of air pollution. Given that 70 percent or more of such coal is burned to produce steam to turn generators for electricity, coal, air pollution, and electric power questions are interrelated. Burning coal causes pollution that both damages the environment esthetically and causes major public health problems. Coal creates carbon dioxide, sulfur dioxide (yellow haze), sulfates (likely the cause of acid rain), particulates (dust), nitrogen oxide, polycyclic hydrocarbons and trace metals that may have carcinogenic effects. Pollution from burning coal is thought to increase the incidence of most lung problems, such as cancer, emphysema, bronchitis, pneumonia, asthma, and colds.[55]

The 1970 Clean Air Act amendments set national standards for sulfur dioxide, nitrogen dioxide, and particulate pollution, which affected the burning of coal, especially for utilities. Enforcement was to be conducted by state governments, which were required to mark off separate air basins and develop a pollution control program for each basin. The national EPA would then review the separate basin plans for approval. Implementation of the plan would be monitored by both state and federal officials. Local air pollution control plans and their implementation became a long process of negotiation among national, state, and local officials and the regulated corporations. Lawsuits were often filed by environmentalist groups impatient with the slow speed of enforcement.[56]

At first the states allowed pollution control through the use of tall stacks to disperse coal emissions. Such utility stacks might be 500 feet high, dispersing emission far above the ground and lowering the pollution measured by regulatory checkpoints on the ground. But this system of pollution control simply increased the amount of emissions blown long distances into someone else's home territory. In the early and middle 1970s, research increasingly indicated that sulfur dioxides, blown into the atmosphere for several days, might be catalyzed by sunlight into sulfates, a constituent of acid rain. In any event, by 1976 the courts placed significant restrictions on tall stacks as a means of pollution control.[57]

Society's rejection of tall stacks for pollution control led to the political battle termed "the scrubber wars" by Walter Rosenbaum.[58] An alternative to tall stacks is the installation of the flue gas scrubber, a large, cumbersome, and expensive mechanism that sprays lime water into the emissions from burning coal, producing a chemical reaction with sulfur dioxide, taking most of it out of the emissions in the form of a solid, or "sludge." Environmentalists and many government officials wanted to require utilities to install scrubbers, but utilities resisted adamantly, arguing that scrubbers would increase the cost of producing electricity by 15 to 20 percent. Furthermore, they pointed out, the equipment was not only expensive, but it broke down frequently.[59]

A third way to control sulfur-dioxide emissions is the use of low-sulfur coal. Depending on the nature of coal seams, coal from one source might have five times more sulfur than coal from a "clean" source. There is plenty of such clean coal in the United States. Accordingly, one can produce 80 percent drops in sulfur emissions simply by burning low-sulfur coal.

But what is technologically obvious may not be politically expedient. Strip-mined coal from the Great Plains is clean, low-sulfur coal. Coal mined in Ohio, Illinois, western Kentucky, and much of Appalachia is high-sulfur coal. If the EPA encouraged the burning of low-sulfur coal as a means of meeting air pollution standards, mines producing high-sulfur coal would close down permanently. Further, electric utilities located near high-sulfur coal, cheaper due to low transportation costs, would see an increase in production costs because coal would now be transported from western sources. Consumers in such areas, especially in Ohio, would pay higher electric rates.[60]

Consequently, a massive lobbying campaign was organized to require the use of scrubbers on all new coal-burning power plants. Such a requirement was part of the Clean Air Act amendments of 1977, the next major national legislation in the air pollution field. The proscrubber coalition was led by congressional politicians concerned about increased unemployment and higher electric rates in their home states—Senator Robert Byrd of West Virginia, the Senate Democrat party leader in 1977, and Ohio's Democratic senators John Glenn and Howard Metzenbaum. Backing such politicians were the owners and miners of high-sulfur coal. Environmentalists also backed the required installation of scrubbers because it might force the expenditure of billions in research by utilities for improved scrubber technologies. Environmentalists also opposed increased coal production in the Great Plains, which would lead to a major increase in strip-mining.[61]

The opposing coalition was much weaker. Its strongest elements were scientists, technologists, and economists committed to the idea that burning low-sulfur coal is the surest and cheapest way to reduce sulfur pollution. EPA officials became convinced of this, but were overruled by the scrubber lobby in the Carter White House.[62] While Derthick and Quirk have shown that "the power of ideas" may persuade politicians to overturn the special interests, as in the case of trucking deregulation, this surely did not happen in the scrubber wars.[63]

The most active interest groups opposing the scrubber requirement were health groups, such as the American Lung Association, that believed the scrubbers were too expensive and impractical to be the basic source of pollution reduction for coal-burning plants. Western coal interests were not strongly represented, for coal producers in the West sometimes owned high-sulfur mines in the East, and strip mines have too few workers to have a major lobbying impact. Western citizens were split between those favoring economic development and those opposing strip-mining, including not only environmentalists, but also ranchers, hunters, recreation industries, and Native Americans. Those utilities that would save money by burning low-sulfur coal without scrubbers did not significantly affect the lobbying battle.[64]

A second major policy battle preceding the NCPP was "prevention of significant deterioration," known as PSD in political jargon. The 1970 Clean Air Act set standards for seven types of pollution. But what if an area already had cleaner air than the standards? Would it be all right to build an electric power plant that would pollute, but no further than the national standards? Congress avoided this issue, which the Sierra Club and other environmental lobbies immediately took to the courts. In a split decision, the U.S. Supreme Court decided that such "significant deterioration" violated the 1970 law, at least in some cases.[65]

By 1975 the legal picture for PSD was not entirely clear, but the situation threatened economic development in Utah and other pristine areas of the West. Federal clean air laws, as interpreted by the courts responding to environmentalist litigation, might prevent the construction of any power plants or factories over a large area. Accordingly, western economic interests launched a campaign to define the PSD concept to permit considerable construction of new projects. Environmentalists were adamantly opposed to such revisions, but the western economic interests, led by Senator Jake Garn of Utah, won a partial victory in the 1977 legislation. New pollution was permitted in PSD areas, but pollution could not reach the national standard. Construction was forbidden in zones around national parks and wilderness areas.[66]

The NCPP met during 1977 as the new round of Clean Air Act amendments requiring scrubbers finally passed Congress with the support of President Carter. Although they had suffered a defeat on the PSD issue, environmental lobbyists were generally satisfied by the air pollution legislation on the books, for they too supported scrubbers.[67] The NCPP sidestepped the scrubber issue, but produced agreement on other issues of implementation of the Clean Air Act.

CONCLUSION

In 1976, as the NCPP was being organized, coal policy issues were much more technically and politically complex than they had been in 1966. The environmental issue had become a priority on the nation's agenda, and coal production and burning was linked to environmental problems in a number of complex ways. An "energy crisis" had seized the nation in 1972, and though concern waned in the middle of the decade, much policy attention was still devoted to developing new energy sources and assessing the role of coal in the overall national energy situation. These issues were now linked to national security issues—to what extent should the nation become energy independent, and to what extent should the United States rely on coal for such independence?

Another complex element was the appearance in the early 1970s of environmental lobbies effective in the politics of coal issues, as well as in the politics of energy and environmental issues linked to coal. Executives of the coal and utility industries had to deal with a new force, and the resulting political and legal battles produced antagonism on all sides. Environmental lobbying was cutting the profits of business, often by initiating long political and judicial battles to block new economic development projects. Needless to say, coal and utility executives were irritated by this strategy.

The new technical and political complexity surrounding coal policy cannot be said to have caused the formation of the NCPP. But neither is it surprising that an experiment in cooperative pluralism was launched in the area of coal policy. The chemical industry uses a great deal of electricity, and some Dow Chemical executives were drawn to questions of energy policy and to the use of coal, which produced more than half of the nation's electric power. It is not surprising that some business executives might wish for a more rational, discussion-oriented mode of formulating coal policy. Such an effort would undoubtedly interest foundations and university institutes with a special concern for energy issues. Meanwhile, certain environmental lobbyists might be intrigued by the prospect of discussing issues with business executives. Accordingly, the NCPP was born.

3

An Organizing Principle: The Rule of Reason

The NCPP was organized by business executives and environmentalists; governmental officials were not invited to participate. As such it constituted an experiment in interest group theory, as normally adversarial groups attempted to change public policy by negotiating among themselves—"groups without government." The organizers of the NCPP were fully aware of its role as an experiment in the potential impact of groups upon public policy, but of course they were not aware of this theoretical interpretation of their activities.

The chief founder of the NCPP was Gerald "Jerry" Decker, who conceived the idea of the NCPP as an outgrowth of his activities as corporate energy manager of the Dow Chemical Company, which ranks near the top in energy use in an industry that requires enormous amounts of electric power. Although Dow Chemical did not adopt a pioneering policy in the business world by refusing to manufacture napalm, as so many anti–Vietnam war demonstrators urged, it may surprise some to learn that in the 1960s and 1970s Dow Chemical was the first Fortune 500 company to address the developing energy problem and to introduce wide-ranging energy conservation techniques within corporate manufacturing processes. Liberal energy writer Daniel Yergin singled out Dow Chemical for praise as a model of corporate energy conservation, Dow having increased the productivity of its energy inputs by 40 percent "with relatively little capital investment" in the period of roughly 1967–1978.[1]

Jerry Decker held an M.S. degree in physics and chemistry from the

University of Michigan and had worked for Dow as a manager since 1940. As energy manager, Decker's duties went beyond the technical tasks of initiating and implementing energy conservation practices: his duty was to communicate within the company the peculiar energy consciousness of Dow's top management, a corporate tradition since the company's founding in the early twentieth century by H. H. Dow, whose hobby was experimenting with electric generating devices. H. H. Dow encouraged managers to introduce novel ways of manufacturing electricity, even if they at first saved no money, and thus Dow Chemical was among the first companies to use electric cogeneration systems in the 1920s. Concern for energy conservation continued within Dow because the company's headquarters plant, located in Midland, Michigan, was forced to introduce energy conservation practices to justify itself as a major center of production in relation to other plants in Texas, which in the early 1950s were using extremely cheap power derived from burning very inexpensive natural gas.[2]

By the late 1960s, Dow's official outlook was to expect energy shortages in the 1970s, and Decker was called upon to communicate this then-deviant outlook to other businesses. When the national government became aware of developing energy shortages—to some degree in 1972, and most acutely after the Arab embargo of October 1973—it was not surprising that Decker received a series of appointments to federal advisory councils on energy policy. In 1975 the Ford administration, under the special direction of Vice-President Nelson Rockefeller, issued the Project Independence report, a plan for attaining national energy self-sufficiency. The Department of Commerce Technical Advisory Board established a panel of advisors on the Project Independence "blueprint" to get the opinions of industrial users of energy, many of whom had complained previously about the impracticality of Project Independence. Decker was appointed codirector of this panel, which emphasized that the potential for increased coal use had been neglected in the Energy Independence report. Decker perhaps went to the panel with this belief, since Dow itself owned coal deposits in Texas, and one might presume that the panel's deliberations strengthened his views on the importance of coal development.

By 1975 Dow management recognized that, for better or for worse, environmentalists and local citizens' groups could organize effectively to create major delays in the development of energy installations. Because

Dow expected higher oil prices, it had encouraged the development of a nuclear power plant at its home base in Midland, Michigan, and by 1975 protest against the nuclear plant had already taught Dow management the difficulties in relying on nuclear power.

> Dow had a most frustrating ordeal in our first effort to purchase power from a new nuclear power facility in Midland, Michigan. The nuclear project became entangled in a bitter, lengthy litigation process that [by 1975] delayed completion by about eight years and saw costs escalated roughly tenfold from the original estimates. The company was very discouraged by the experience and felt that a better way could be found to work out its energy strategy.[3]

Dow management in 1975 foresaw that coal development would encounter some of the political resistance that nuclear energy had experienced. Yet Dow wanted coal to be developed as a national policy because (1) this would prevent nationwide energy shortages, (2) Dow itself owned coal, and (3) "Dow scientists were telling us that coal will be the raw material source of the future for the chemical industry." According to a Dow executive, Decker and his immediate associates "concluded that the most important thing to be done from the energy users' standpoint was to search for a process that would help to prevent the kind of delays in our coal development program that had been encountered in the nuclear project."[4]

To continue in the words of Macauley "Mac" Whiting, an associate of Decker's at Dow who succeeded Decker as the chair of the industry caucus of the NCPP:

> It was at this point that we broadened our group within the company to others who had experienced delays for environmental reasons in the manufacture of chemicals and other activities. They mentioned that a process called the "Rule of Reason" had been used in much smaller situations and had been successful in getting the disputing parties beyond the emotional point to a rational discussion of the issues . . . It was concluded that consideration of the pros and cons in a relatively unemotional setting was a promising technique to apply to coal related issues.[5]

In other words, Dow's corporate litigation department had a pioneering interest in a method of alternative dispute resolution that could serve as a substitute for the normal adversarial process of litigation. This particular process, the Rule of Reason, was modeled on aspects of antitrust law and had been conceived by Milton R. Wessel. In fact, at the very time that Decker, Whiting, and others were discussing policy toward coal development, Dow's General Counsel William A. Groening, Jr., described the new technique in the foreword to Wessel's book.[6] Perhaps it was fortuitous that Dow Chemical was both in the forefront of corporate consciousness about energy and similarly advanced in its interest in alternatives to traditional litigation. One might speculate, however, that the attack on Dow Chemical in the 1960s for its manufacture of napalm rendered management unusually sensitive to maintaining a good corporate image. As Groening wrote in the foreword to *The Rule of Reason*: "Not only must litigation be conducted in such a way as to protect the best interests of the corporation—in short, to win—but while the litigation is pending the corporation must show an open and honest face, one that will also enable it to win in that most important court of public opinion."[7]

In any event, one circle of Dow's management defined a problem (the need to balance development and environmental interests in the coal sector), while another circle proposed a solution: use the Rule of Reason among the environmental and corporate disputants. When the problem and the solution united in discussions among Dow executives, the fundamental conception of the NCPP came to exist.[8]

THE RULE OF REASON

The NCPP was an unusual negotiating group, for it had a "bible," a book to provide a common doctrine and a code of behavior for the group. *The Rule of Reason: A New Approach to Corporate Litigation*, written by Milton R. Wessel, a New York trial lawyer and law professor at New York University, was a common reference point and it influenced the distinctive "culture" and values of the NCPP, and the norms of interaction among the participants. The book promoted the surprising success of the negotiators in reaching agreement. Indeed, virtually every time in which NCPP leaders were asked to describe the project, they referred to *The Rule of Reason* as a distinctive aspect of the NCPP and a major reason for its success in reaching agreements. For instance, *Where We Agree:*

Summary and Synthesis, the most widely distributed NCPP report, states:

> The National Coal Policy Project emphasized reaching agreement, where possible, rather than seeking victories. To facilitate this effort, the project adopted a set of negotiating principles known as the "Rule of Reason." These principles . . . include the following:
>
> - Data should not be withheld from the other side.
> - Delaying tactics should not be used.
> - Tactics should not be used to mislead.
> - Motives should not be impugned lightly.
> - Dogmatism should be avoided.
> - Extremism should be countered forcefully . . . but not in kind.
> - Integrity should be given first priority.
>
> Agreement to use these principles helped convince participants that the project could resolve some of their differences constructively, and as it turned out, conducting project meetings in the spirit of the Rule of Reason did facilitate the search for workable solutions to the difficult issues being addressed.[9]

As the quotation indicates, the Rule of Reason can be succinctly stated as a seven-point behavioral code, which is easily understood and remembered. The code was reinforced by Father Francis X. Quinn, S.J., who presided over the plenary meetings, and by the chairs of the various subcommittees and task forces. Father Quinn referred to the Rule of Reason in prayerlike invocations before the business of the plenary meeting. A social phenomenon occurred similar to those reported by anthropologists: the Rule of Reason acquired a symbolic importance and received a certain veneration. It provided a common culture, a code of behavior for the group. Behavior threatening the success of negotiations was criticized by committee leaders as violating the Rule of Reason.

The Rule of Reason, then, helped attract participants. It could be touted as a new and promising method to convince adversaries to become more agreeable. The NCPP's participants became more optimistic about the prospect of deriving success from inevitably arduous discussions, because they possessed a method of communication ordinarily lacking in the judicial and legislative processes. And because the Rule of Reason became an element of group culture, it helped to bind the group together.

The book *The Rule of Reason* had three layers of meaning for the NCPP. First, it provided a short and readily understandable set of rules for nonadversarial behavior. Second, it provided a longer list of nineteen rules, incorporating the short list, to further elucidate nonadversarial behavior. Nineteen rules of behavior, sanctioned by a group leader and regarded as legitimate by the group, are probably enough to change dramatically the behavior of any committee. However, these rules are only one part of one chapter of a 221-page book, which provides a theory to back up the rules.

The book has a somewhat unusual structure: it is written by an idealistic Machiavelli advising his Prince. Wessel directs the book specifically to corporate general counsels and advises them that social changes require changes in their legal practices, summarized in the term the Rule of Reason. As a corporate lawyer advising other corporate lawyers, Wessel is keenly aware that corporate policies can affect the welfare of society as a whole in such matters as chemical hazards, but he certainly does not identify himself with the public interest lawyers: "I use with reluctance the terms 'public interest group,' 'environmentalist,' 'consumer advocate,' and even 'government' to describe industry's litigating adversaries . . . for these groups no more speak for the community as a whole than does industry."[10]

But after absorbing this and a couple of other jabs, public interest litigators certainly could read Wessel's book with interest. Wessel argues that because of the public interest movement and its effect on legislation, and because of technological change, businesses are now engaged in a new type of litigation that he styles "socioscientific," because it involves all of society, and is characterized by complex, scientific issues. Although Wessel has no general objection to adversarial processes of litigation, and himself had been a trial lawyer for twenty-five years, he believes that adversarial tactics are no longer well suited to the socioscientific cases dealing with environmental, product safety, occupational health, civil rights, and antitrust issues. In particular, Wessel objects to the application of "sporting tactics," or legal gamesmanship, to socioscientific litigation. Legal gamesmanship, which treats the handling of facts as part of legal strategy, does not pose insuperable barriers to justice in traditional civil torts litigation, in his view. There, the competition among combatant lawyers usually serves to bring out the facts. But socioscientific litigation is different in three respects. First, unlike traditional cases, the most important interests, those of society as a whole, are not represented by legal counsel. (Wessel does not regard public interest lawyers as ordinarily representing the general interest.) "Second, and critical to the decision process, there are no absolute answers [in socioscientific disputes]. Ultimately, the decision must depend on how one values one set of concerns (e.g., energy), as against another (e.g., pollution)."[11] "Third, . . . much of the evidence to be evaluated is of the most complicated scientific character, largely incomprehensible even to the intelligent layperson."[12] Wessel was particularly upset by instances of corporate destruction of data as part of out-of-court settlements.[13] On the other hand, Wessel was upset at "so much environmental and consumer extremism" leading to the "interminable claims and suits" that motivated corporations to hide data.[14]

Wessel implies that nearly incomprehensible cases involving complex shadings of right and wrong related to scientific data, and in which the general interest is poorly represented, are not likely to be resolved with justice unless new procedures to gain comprehensible information are developed, procedures such as the Rule of Reason. Writing at a time of especially pervasive criticism of corporate behavior, Wessel believed that corporations should take a unilateral initiative and present openly most of the technical data relevant to scientific disputes. This would make a favorable impression on judges and juries who normally seemed hostile to the corporate point of view in environmental and consumer litigation. Such open presentation of data would also make better public policy, while environmentalist litigants would soon realize that their own interests would normally lie in similarly open presentation of data.

This theoretical discussion and its refinement into nineteen rules of behavior appear in the book's first chapter. Succeeding chapters demonstrate applications of the Rule of Reason approach to the stages of litigation: preliminary preparation, settlement, pretrial preparation, the conduct of hearings and trials. Probably the majority of participants in the NCPP read no more than the first chapter of Wessel's book. Still, that chapter covers considerable ground, and includes the list of nineteen rules that constitute an application of the Rule of Reason to the negotiation process.

Besides providing short and long lists of negotiation rules and a theory from which they were derived, *The Rule of Reason* actually integrates this theory with a statement explaining why society should develop nonadversarial processes. In other words, the book integrates the short list of rules, the long list of rules, the theory of why openness is preferable in socioscientific litigation, and the explanation of why business and society need such new norms to handle disputes. This synthesis constitutes an impressive combined system of theory and practice.

Further, *The Rule of Reason* impresses both business people and environmentalists for reasons beyond its inherent intellectual attractiveness. The book is written by a business lawyer for other business lawyers, and thus, not surprisingly, it appeals to those with a corporate point of view. It appeals to environmentalists somewhat less, yet Wessel accepts much of the point of view of environmentalism and clearly seeks a better method to represent environmental values in litigation. He admits that public values are often neglected in socioscientific lawsuits. Thus the overall effect of the book was particularly useful to the NCPP. Wessel's concern for reason, moderation, integrity, and the public interest provide an example of the sort of businessperson an environmentalist might fruitfully negotiate with.

During the twelve hours of NCPP plenum and committee meetings I observed, no instances of hostile interaction occurred. Discussion proceeded in a factual manner, and participants were sensitive to the need to make trade-offs between environmental and developmental values. This respectable decorum provided some corroboration for the repeated statements by almost all NCPP leaders in interviews and in project publications that acceptance of the Rule of Reason procedures had made a major difference in the behavior of the participants. The amount of interpersonal conflict within the NCPP seems to have been surprisingly low. The transportation committee did disagree about the scope of its report, but this difference was not accompanied by strong conflict among the committee members. One example of personality conflict was remembered several years later, however. In this instance, staff director Frank Murray believed that the Rule of Reason, combined with a structural change in the pricing committee, alleviated the problem. Murray described the situation to researchers Tina Hay and Barbara Gray as follows:

> The belligerent, dogmatic behavior of an industry participant had forestalled progress in the Pricing Task Force. The industry caucus determined that their member felt intimidated by the stature and ex-

> pertise of the two Ph.D. economists from the environmental side. Frustrated at his inability to counteract the environmental side's theoretical arguments, the industry member had resorted to power tactics. The caucus located a Ph.D. economist to serve as an adviser to the industry members of the task force, thus relieving some of the pressure on the problematic member. The caucus also convincingly persuaded the member to abide by the Rule of Reason. Indeed, one prominent caucus member told the offender quite bluntly to either change his behavior or resign the project.[15]

Murray and others remembered that participants needed to be reminded of the Rule of Reason only occasionally.[16]

The Rule of Reason (19 points)

- Data will not be withheld because "negative" or "unhelpful."
- Concealment will not be practiced for concealment's sake.
- Delay will not be employed as a tactic to avoid an undesired result.
- Unfair "tricks" designed to mislead will not be employed to win a struggle.
- Borderline ethical disingenuity will not be practiced.
- Motivation of adversaries will not be unnecessarily or lightly impugned.
- An opponent's personal habits and characteristics will not be questioned unless relevant.
- Wherever possible, opportunity will be left for an opponent's orderly retreat and "exit with honor."
- Extremism may be countered forcefully and with emotionalism where justified, but will not be fought or matched with extremism.
- Dogmatism will be avoided.
- Complex concepts will be simplified as much as possible so as to achieve maximum communication and lay understanding.
- Effort will be made to identify and isolate subjective considerations involved in reaching a technical conclusion.
- Relevant data will be disclosed when ready for analysis and peer review—even to an extremist opposition and without legal obligation.

- Socially desirable professional disclosure will not be postponed for tactical advantage.
- Hypothesis, uncertainty, and inadequate knowledge will be stated affirmatively—not conceded only reluctantly or under pressure.
- Unjustified assumption and off-the-cuff comment will be avoided.
- Interest in an outcome, relationship to a proponent, and bias, prejudice, and proclivity of any kind will be disclosed voluntarily and as a matter of course.
- Research and investigation will be conducted appropriate to the problem involved. Although the precise extent of that effort will vary with the nature of the issues, it will be consistent with stated overall responsibility to solution of the problem.
- Integrity will always be given first priority.

BARGAINING AND NEGOTIATIONS THEORY

Wessel wrote *The Rule of Reason* from the standpoint of a practicing attorney. But since 1960 a social science theory of bargaining and negotiations behavior has appeared. In 1982, Zartman and Berman found more than a thousand research reports published in this field by psychologists, sociologists, political scientists, and economists.[18] Thomas Schelling, Howard Raiffa, and other economists have constructed formal models of conflict and conflict resolution that have had a major impact outside of the discipline of economics.[19] The social science research regarding bargaining and negotiations sometimes appears summarized in popular works, the most famous of which is *Getting to Yes* by Roger Fisher and William Ury, a widely distributed book that attempts to instruct the layperson how to negotiate common types of disputes and that is heavily influenced by the pioneering research by Harvard economist Raiffa.[20] Wessel's specific injunctions to negotiators closely resemble the instructions for finding common interests in *Getting to Yes*. Accordingly, Wessel's independent thought on the subject and its usefulness in gaining agreement in the NCPP further validates the more strictly academic theory.

Fisher and Ury summarize their ideas for finding mutual interests as four generalizations: (1) separate the people from the problem; (2) focus on interests, not positions; (3) invent options for mutual gain; (4) insist on using objective criteria.[21]

Wessel of course emphasizes the need to avoid personal attacks in the negotiation process. (This is common sense, but emotional disputants often forget this. Consider divorce negotiations.) Similarly, Wessel's ideas stress the need to understand the interests of the other side, to search for a probable substantial degree of common interests, and to alter one's position flexibly to attain these mutual interests.

Wessel has less to say about inventing "options for mutual gain," because he believes that such options will be found in objective data revealed by the Rule of Reason process. He does, however, encourage corporate counsel to search for such options during all the phases of the litigation process.[22] In the actual practice of the NCPP, however, its leaders actively invented options for mutual gain in applying basic market models to public policy.

Wessel's approach also overlaps with Fisher and Ury's fourth generalization: "insist on using objective criteria." The authors state: "The more you bring standards of fairness, efficiency, or scientific merit to bear on your particular problem, the more likely you are to produce a final package that is wise and fair."[23] This is Wessel's core argument. Fisher and Ury develop their argument by citing a case which also exemplifies Wessel's injunctions: during a conference on the law of the sea, an impasse was broken by an MIT model of the economics of deep-seabed mining. Wessel's point is that in American civil litigation, if one side performed economic analysis but the results did not support that side's bargaining position, the litigator would conceal the information from everyone.[24] Fisher and Ury conclude a chapter with injunctions that recall Wessel's outlook: "Frame each issue as a joint search for objective criteria"[25]; ask one another, "What's your theory?"; and "Reason and be open to reason."[26] Wessel provides a set of rules for negotiation processes, particularly those involving technical disputes over scientific issues. He does not, however, claim to present a set of rules for *all* negotiations, but primarily for those involved in "socioscientific" litigation. Accordingly, much of his work advises corporate counsel how to conduct litigation under the Rule of Reason. In this sense, Wessel is not a general theorist of bargaining and negotiation.

THE ADVERSARIAL PROCESS

The founders of the NCPP in the Dow Chemical Company believed that the adversarial processes of ordinary lobbying and litigation neglected the discovery of common interests. Thus they became interested in testing Wessel's Rule of Reason discussion process. In addition, some environmentalist leaders were critical of the normal adversarial decision-making processes. Such was the case for some participants in the NCPP, such as J. Michael McCloskey, Executive Director of the Sierra Club, who stated:

> Both sides have tended to assume that the issues have all been non-negotiable. Industry has tended to assume that environmentalists oppose everything, while environmentalists have tended to assume that industry will never make any real concessions. With these attitudes many disputes instantly become ideological.
>
> This is unfortunate because many environmental disputes involve practical differences which are resolvable. My own experience suggests that half or more of all disputes over site-specific construction projects really involve only the limited aim of gaining mitigation.
>
> The polemics of the dispute, however, make it difficult for the parties to reveal their ultimate aims and their final "bottom-line" in any settlement. Thus it is easy for the two sides to misread each other's motives and intentions.[27]

Participants in the NCPP had read *The Rule of Reason*, or at least were familiar with its critique of the usefulness of standard litigation procedures, and its leaders frequently articulated a reasonably clear position on the defects of the "adversarial system" in these situations. As the staff director and his assistant wrote:

> When speaking of the adversary system, we mean those functions of the courts, legislatures, and administrative agencies (and their subordinate bodies) that exercise public policy, decision-making authority . . . The essence of the traditional system is that parties are pitted against one another in a variety of ways in different forums and must generally prove that their rights are superior to the rights of the opposing parties. The parties present their case to a decision maker who is not a party to the dispute. To convince the decision maker of

> the correctness of their claim, the parties often use various tactics to make the opponent's claim appear as weak as possible. In practical terms, this eliminates the middle ground as a source of agreement, acceptable to all involved.[28]

The two staff directors, Murray and Curran, contrast this system to the "basic thrust of the NCPP . . . to encourage the parties to deal directly with each other in attempting to reach agreements . . . because there was no third party present." The result was "to allow the parties to reveal their real concerns and objectives, to express doubts or lack of knowledge, to change their minds, and, most importantly, to search for reasonable, workable solutions."[29] When the Rule of Reason process works, the result is an ability to communicate subtle and complex information about facts and opinions, which can be more useful than the traditional mode of "one expert's opinion . . . countered with equally qualified opinion again and again." Such polarized communication causes "excessive delay and cost" that are "frequent concomitants to the use of the adversary system." "This is especially true when the case in point involves issues concerning the public interest" that are characterized by "the complexity of the issues, whether they are defined as social, economic, environmental. . . . Such issues often demand decisions other than the win-lose type that the [adversarial] system is best able to produce." Decision making concerning complex cases involving public interests "requires creativity and flexibility, qualities that the parties can prevent . . . if they rely on a strictly adversarial approach."[30] In the words of David Abshire, director of the Georgetown University research institute which sponsored the NCPP, "The emerging issues are much too complex and subtle, and the adversarial system is too divisive, to deal with the vital issues that we see looming before us."[31]

In short, the nonadversarial process of the Rule of Reason was designed to bring together disputants to communicate directly and to resolve nonfundamental conflicts more quickly through processes of sharing complex technical issues involving questions of the public interest. Such a process of rich communication might enable disputants to find unsuspected middle ground.

After they were criticized in 1977 for attempting to substitute an elitist negotiating process for more representative institutions (see below), Frank Murray and other leaders of the NCPP always added in public statements that their goal was not to change the whole American system of government. "The adversary process follows directly from the Constitution and is the foundation of a system that has served our nation well over the past 200 years."[32] "We are not proposing that the process of discussion and negotiation in which we have participated should replace the adversary process."[33] "In cases where the rights of parties can be determined by reference to contracts or specific laws, this process works reasonably well. For many emerging socioeconomic issues, however, the traditional process is fraught with difficulties."[34]

CONCLUSION

Communication among adversaries, needed for cooperative pluralism, can be enhanced by the development of special cultures of communication. The Rule of Reason is one such system of communication; probably others could be developed if cooperative pluralist institutions were to become numerous. The Rule of Reason enhances citizenship, as Mill and Muir define it, for participants are concerned to treat one another as equals in a learning process in which they educate one another about public issues. Muir envisions an ideal legislature in which members sometimes oppose one another and sometimes educate one another. The NCPP, guided by the Rule of Reason, was intended to provide a similar institution for leaders of interest groups.

Wessel's *The Rule of Reason* is not to be viewed as providing a sure recipe for successful negotiations. Its effectiveness, as demonstrated by the NCPP, is probably due to Wessel's independent realization of many of the concepts developed by social scientific scholars of bargaining-negotiations theory. Accordingly, as this theory continues to develop, and as new modes of application are tested in actual negotiations, the resulting codes for successful negotiations will be more sophisticated than Wessel's method.

4

Early History of the National Coal Policy Project

In his role as corporate energy ambassador, Dow's Jerry Decker had discussed the need to develop coal resources with 200 to 300 customers, many of whom agreed with him and urged Dow Chemical to take leadership in resolving developmental and environmental interests. Decker and Whiting discussed the Rule of Reason approach with federal government officials and were encouraged to proceed:

> We talked with people from ERDA [Energy Research and Development Agency, a predecessor to the Department of Energy], the Department of Commerce, FEA (Federal Energy Agency, another predecessor to the Dept. of Energy], and the Department of the Interior. In all cases we received encouragement and support from these people. Still, they too were skeptical whether the United States was ready for a large group of industry people and a large group of environmental people to attempt to negotiate reasoned conclusions on coal issues.[1]

Decker and Whiting then made a mistake that might have killed the idea of launching a Rule of Reason process on coal issues. After being encouraged by government officials, they took the straightforward approach of organizing a meeting with some environmentalist leaders to discuss their idea. Decker assembled about a dozen environmentalists who had served with him on the FEA's Environmental Advisory Committee. The result was a disaster: the meeting had no structure, and the imbalance of numbers between the environmentalists and Decker and Whiting promoted an atmosphere in which the environmentalists attacked the Dow executives. In the words of Mac Whiting:

> The next step was to face environmental "lions" in their den . . . This meeting was one of the more humiliating moments of our careers for Jerry and me. I must say, in all candor, that we did not succeed in winning a lot of friends or influencing a lot of people in the three or four hour ordeal that afternoon . . . I am sure that we "lost" that one . . . everyone was candid . . . At that time, the environmentalists did not seem receptive to the idea of resolving differences via the Rule of Reason.[2]

The problem seemed to be one of structure—imbalance inducing one side to attack the numerically weak side—rather than a true lack of interest in the Rule of Reason process. After corporate lambs Decker and Whiting had been torn to shreds and resurrected in Midland, they called the twelve environmentalists individually, and "somewhat to our surprise, there was a guardedly favorable response."[3] One of those present at the disaster meeting, Laurence I. "Larry" Moss, was especially interested in the concept and provided leadership for environmentalists willing to try the experiment. Moss eventually became coleader of the NCPP with Decker, and later with Whiting, after Decker resigned to become energy manager for Kaiser Aluminum in 1978.

The next stage of the organization was successful. Decker, Whiting, and Moss were able to fund an experimental meeting with grants from the Ford and the Rockefeller foundations. To take their enterprise out of Dow Chemical and to find a neutral ground, they persuaded an academic research institution, Georgetown University's Center for Strategic and International Studies (SIS), in Washington, D.C., to become the administering institution for the fledgling Coal Policy Project.

Ford Foundation officials found the Rule of Reason idea creative, unusual, even slightly bizarre, but so compelling that they could not turn away from it, especially as the Ford Foundation's funding activities were indirectly responsible for a great deal of environmentalist litigation in the first place. David M. Abshire, the head of Georgetown University's CSIS, had a similar reaction to the idea, remembering it as "a novel project that many thought could never work . . . not normally appropriate for an institution such as ours . . . [yet] it was too important to walk away from."[4] Abshire and the director for energy studies of CSIS, Francis X. Murray, both became strong believers in the value of the NCPP experiment and were willing to incur temporary financial losses on behalf of the project, although CSIS eventually raised $1.4 million and the books were balanced.

The next step was to organize, in July 1976, a pilot meeting to convince possible participants and funding sources that the Rule of Reason process might work. The pilot meeting was carefully structured, with the earlier disaster well in mind. About twenty-five people attended, representing a roughly equal split between businessmen and environmentalists. The famous labor mediator, Harvard professor, and former secretary of labor John Dunlop was hired to chair the process. This time the Rule of Reason discussion process worked. According to Mac Whiting:

> Two issues were discussed at that meeting: energy pricing, and the prevention of significant deterioration of air quality. There were many surprises for all of us. For example, the industry people were surprised to learn that most environmentalists were proponents of higher energy prices. [At least this was true of those participating in NCPP functions.] To the industrialists, that was not the point of view we expected. The fact that there was such a recognition of economic laws in the environmental movement was truly a great surprise to the industry representatives. I think Larry Moss would agree that there were also some favorable responses for the environmentalists.[5]

Under the aegis of John Dunlop, both sides engaged in reasoned discussion about relatively specific issues without recrimination: environmentalists were willing to talk about costs and benefits; businessmen indicated a belief in the legitimacy of considering environmental concerns in corporate decision making. In Whiting's words, "The result of the meeting was a decision to go ahead with the project."[6]

Decker and Moss decided to convene a two-sided negotiation process (see below) with an industry side, Decker as chairman, and an environmentalist side, Moss as chairman. The first general meeting of the NCPP was convened six months later, January 18–19, 1977, in a Washington suburb with personnel recruited primarily by these two men (see below). During these months, fund-raising was difficult, but Abshire and the CSIS decided to stay with the project and hope it could break even financially.

The origin of the National Coal Policy Project is here described largely as a result of the unusual history and management traditions of Dow Chemical. Leaders of the NCPP normally did not describe its history this way, perhaps because they generally strove to give a "bipartisan" description of their organization as a business-environmentalist hybrid. Certainly I do not consider the NCPP to be a one-sided probusiness group that co-opted environmentalist participants. The reader may judge for her- or himself while inspecting the policy positions of the NCPP in succeeding chapters.

Still, if Dow Chemical's own chemistry of foresight about energy and unusual interest in negotiating with environmentalists provided the raw materials for the NCPP idea, the project probably would not have crystallized without the efforts of Jerry Decker, who might be regarded as the "chemist." (Political scientists would refer to Decker as an organizational entrepreneur.)[7] Decker's personal dedication to a Rule of Reason process in determining coal policy is indicated by the fact that, even after being humiliated by the criticism of hostile environmentalists, he persisted, telephoning each environmentalist individually, surely not an easy thing for him to do, but which was crucial to the development of the NCPP. Decker conveyed an impression of trustworthiness, dedication, and sincere idealism—a bit naive, perhaps, but someone who sincerely believed that adversaries could cooperate after studying the facts and holding face-to-face discussions. Virtually no one believed Decker to be on an "ego-trip," as the saying goes. Decker was thus persuasive to the Ford Foundation, to CSIS, and to Larry Moss and other environmentalists.

COAL AND ENERGY INDEPENDENCE

Some of the organizational interests and personal motives leading to support for the NCPP are mentioned below. But the project had definite goals set forth by its founders that we should consider. The objectives of the NCPP were basically twofold: (1) to indicate how America might become more self-sufficient in energy through a program to develop coal and how it might balance such development by incorporating environmental objectives; and (2) to demonstrate the effectiveness of the Rule of Reason discussion process as a tool of gaining agreement between business and environmentalists. These objectives were articulated by Decker and his associates during the initial discussions at Dow Chemical.

The first objective—the NCPP as a tool to gain energy independence—was an important part of the rhetoric of the NCPP, and there is little reason to doubt the sincerity of Decker, Moss, Abshire, Murray, and the Ford and Rockefeller foundations in adhering to this goal. In the final report of the NCPP, immediately after the heading "The Need for the Project," Francis "Frank" Murray states:

> Before the oil crisis of 1973, it was already apparent that the United States was dangerously dependent upon imported oil. The embargo experience, and the immediate impact that it had in this country, heightened our national desire to reduce imports and rely more heavily on domestic energy sources. At that time, the increased utilization of coal was being considered as a means to avoid placing an even greater demand on our depleting oil and natural gas resources.[8]

We should recall that American oil imports as a share of consumption steadily increased, even after the oil embargo of 1973, from about 35 percent at that time to as high as 48 percent in 1978. In 1976, most people seriously concerned about America's energy situation were frustrated at the steady trend of increasing imports and felt anxious about what course of action to take. To these people, increasing coal production seemed important to prevent economic blackmail of the United States by OPEC. As Murray continues in the NCPP *Final Report*, "Frequent calls were heard to substantially increase total national coal production. The doubling of the 1977 production total by 1985 was a central goal of the Carter administration's "National Energy Plan." In short, coal was often expected both to replace the use of imported oil and gas and to provide energy for continued industrial growth in the United States."[9]

This reminds us that the Carter administration came to power in January 1977, and even as the NCPP deliberated in 1977, the official policy of the United States called for American coal production to double in eight years. Few were more enthusiastic than Carter administration leaders about the need to increase coal production in the interests of meeting

America's energy needs and of preserving national security from OPEC manipulation. However, the legal complexities of a rapid development of coal resources were apparent, leading the Carter administration to propose a federal Energy Mobilization Board, which would have had wide power to overrule other federal agencies and perhaps state and local governments in cases of refusals to grant permits legally necessary for coal development. Congress killed the Energy Mobilization Board, however, within a year because of its unpopular centralization of authority.[10]

The NCPP presented itself as an alternative to such centralized authority and to the traditional "adversarial system" of litigation, politicking, and hearings board procedures. The *Final Report* stated: "It became increasingly clear that despite the great promise that coal holds . . . progress would not be made without some resolution of the conflicts in national goals and priorities that would be created."[11] The hope was that the NCPP's process would provide a breakthrough so that the nation could develop coal, protect environmental concerns, and reach agreements to do so with rapidity. Or as David Abshire put it, somewhat more dramatically:

> Energy and environmental issues had collided on the center stage of our political consciousness. As that battle continued it became more shrill and less rational, making the task of thoughtful policy formulation nearly impossible. To continue along this course held great peril both substantively and procedurally for the United States not only for our domestic policies but also for our capability to deal with international challenges. Solutions had to be found. The global energy picture was dark and threatening. New methods and processes for addressing the issues had to be developed.[12]

The director of CSIS even believed the NCPP might have global importance. "Thus the project, despite the limited objective professed by its founders, had much broader implications; if successful it would mark the beginning of a constructive process for resolving these terribly divisive issues and would also make a major breakthrough in the energy fields."[13]

One reason for an increasing interest in the development of coal resources was the realization by some in 1976 that the prospects for nuclear power were not as bright as had been generally hoped a few years earlier. As we have seen, the difficulties of constructing the Midland nuclear plant were an important reason for Dow Chemical's increasing concern for coal development. Other executives in intensive energy-using industries were reluctantly concluding by 1976 that the nation could not simply phase out burning oil and natural gas to generate electricity by rapidly developing nuclear power-plant capacity. This point of view, accepted by almost everyone a few years later, was implicit in the organizational credo of the NCPP.

COAL ISSUES

In regard to policy-making processes concerning coal-related issues, the NCPP's leaders perceived that there had been an exceptional degree of hostility between business and environmentalists in coal politics, and that these tensions were serving no useful purpose. Their statements tended to collapse a discussion of coal issues into the other two categories of dealing with a worldwide energy crisis and of developing procedural alternatives to adversarial institutions. But analytically, problems of coal policy-making per se can be separated from the other two areas.

The founders of the NCPP themselves referred rarely to problems of conflict in coal policy-making, because such comments would be criticisms of some of the people and organizations they wanted to participate in the NCPP. But the coal industry has a public image as conflict-ridden because of the labor history of deep mining (as opposed to strip-mining). Sociologist S. M. Lipset has shown that in industrialized nations labor militancy is often associated with a work force living in geographically isolated communities, having populations that are relatively homogeneous and working in the same industry.[14] Accordingly, miners, sailors, fishermen, and lumberjacks tend to develop a community basis for militant unionism. So, too, at times, have American coal miners, who have had the frequent misfortune of confronting particularly tough, unyielding, and mean-spirited corporate adversaries. The owners and managers of large mining companies do not live near mining communities, and such geographical separation objectifies the worker in the view of management and promotes an unyielding attitude in labor negotiations.[15] On the other hand, many small mines in Appalachia are owned by small businessmen, who live nearby but are unsympathetic to miners' demands as undercutting small profit margins in what has been historically a highly competitive industry.

Coal mining has been the source of class warfare in a literal sense, as in the case of the Molly Maguires (a group of Irish miners using guerilla tactics) in the eastern Pennsylvania coal fields, or the notorious "Bloody Harlan" conflicts in southeastern Kentucky in the 1930s. The archetypal mine union leader, John L. Lewis, had arranged a truce in the 1950s and 1960s, based on corporate contributions to miners' pension funds and hospitals in return for miners accepting the reduction of mining jobs due to the lessening demand for coal and the introduction of strip-mining, which substituted heavy machinery for workers. But after John L. Lewis resigned from the UMW presidency in 1960, no single authoritative figure could long dominate the union, leading to bitter factional struggles, including the notorious execution-style killing of Jock Yablonski, a leader of a rebel faction opposed to Lewis's immediate successor, W. A. "Tony" Boyle, who was subsequently sentenced to the federal penitentiary for planning this crime. Coal miners became militant once again in the middle 1970s, leading to a 111-day strike from December 1977 to March 1978, and to another long strike in the spring of 1981.

As the NCPP was being organized in late 1976, the UMW was undergoing a bitter three-way factional struggle for the control of the union. Organizers of the NCPP were not sure environmentalists and businessmen could agree, and they felt that inviting mine union leaders to participate in a three-cornered negotiating group would clinch the failure of the effort, especially during a time of a hard-fought election in the union when representatives of the three factions could not endanger the popularity of their candidates by compromising worker demands in a negotiating forum, however insubstantial. On the other hand, the NCPP did not seek to address core worker issues of mine safety and health in the absence of union participants.

To my knowledge, no one was killed or critically wounded during the protracted legislative struggle that resulted in the enactment of the Surface Mining Control and Reclamation Act of 1977. However, Louise Dunlap, head of the Environmental Policy Center and lead lobbyist for the SMCRA (pronounced "smack-rah" in political jargon), told me that some of her supporters in mining areas were shot at. This violence perhaps symbolizes the intensity of the seven-year legislative battle leading to the enactment of strip-mining control legislation.

The conflict was exacerbated by the fact that many legislators viewed strip-mining legislation as a matter of ideology. Conservatives usually opposed strip-mining controls as a major extension of the powers of the federal government and another unwanted regulation of private business. Liberals, on the other hand, often saw strip-mining as yet another example of the depredations of business in a selfish struggle to increase profits. It can be mathematically demonstrated that many members of Congress voted against the economic interests of their districts due to their personal ideological position regarding the SMCRA legislation.[16]

A second bundle of issues addressed by the NCPP were related to air pollution from burning coal—not an area in which most of the founders of the project had much experience (with the exception of Larry Moss, who in his Sierra Club role had helped launch a basic lawsuit in air pollution policy).[17] But Moss and the others understood that air pollution is an area of policy in which Rule of Reason discussions had exceptional potential.

Air pollution policy is characterized by the confusion and blundering of leading participants, through no fault of the businesses and environmentalist lobbies involved. Air pollution policy in the 1970s was a new area of public action, and basic scientific principles were in doubt. For instance, at the beginning of the decade, scientists felt that lowering sulfur dioxide (SO_2) emissions was a top priority; at the end of the decade, the trend of scientific opinion changed course, and most scientists held that sulfates (SO_4) and nitrous oxides (NO_2) were more damaging to health than sulfur dioxide. Most readers are familiar with the idea that acid rain is a result of SO_4 emissions being blown about in the upper atmosphere for several days, where they combine with water and, catalyzed by sunlight, form sulfuric acid compounds (H_2SO_4) that mix with rainfall to form acid rain. But this hypothesis has not been definitively confirmed, and uncertainty persists to this day.

To get some impression of the confusion in the air pollution policy area, the reader might turn to such studies of the subject as *Clean Coal/Dirty Air*, by Bruce Ackerman and William Hassler,[18] or *Regulation and the Courts: The Case of the Clean Air Act*, by R. Shep Melnick.[19] Confusion is a central theme of both books. Ackerman and Hassler demonstrate that mandating the installation of expensive scrubbers was probably unnecessary because electric power plants could have reduced sulfur in emissions simply by buying low-sulfur coal. They conclude that better policy could have been made in this instance if Congress had set a clear agenda of goals for the EPA, and had let experts clarify the issues of regulating coal emissions, rather than Congress itself specifying the technical means (e.g. scrubbers) to the end of reducing air pollution. The book clearly portrays the misunderstanding of basic issues by important decision makers, and demonstrates how, meanwhile, special interests, seeking to protect jobs and profits from mines producing high-sulfur coal, managed to achieve their goals by mandating the scrubbing machinery and preserving their ability to compete with producers of low-sulfur coal.

Similarly, R. Shep Melnick argues that frequent judicial action in the realm of air pollution policy led to inappropriate and inefficient policies. The courts were not qualified to decide complex issues of air pollution control (neither was Congress, which was part of the problem). Accordingly, the courts fell back on legal doctrines or dubious interpretations of congressional intent—legal actions tending to be far removed from the chemistry, engineering, and economics of pollution control.

Air pollution policy thus provides fertile ground for productive Rule of Reason discussions between business and environmentalists. Melnick's careful research indicates that air pollution is one area of public policy where the adversarial processes of litigation may further confuse the search for policy that promotes the general welfare. As noted above, the Rule of Reason process stresses the need to be forthright with factual knowledge, rather than concealing knowledge from one's opponents and distorting information to present one's case in the best possible light.

In general, then, air pollution policy is fraught with confusion because of the uncertainty of the scientific knowledge involved. Adversarial processes of decision making add to the confusion by disrupting scientific and political discussions based on a communication of facts and differing values. This situation was apparent to the founders of the NCPP and provided an additional rationale for pursuing a Rule of Reason process in the coal-related air pollution policy area. Intensity of conflict in the strip-mining policy area and intellectual confusion in the air pollution control area provided a justification for a new approach to decision making.

SCOPE, STRUCTURE, AND PARTICIPANTS

Early decisions by Decker, Moss, Whiting, and Murray defined the scope of the task of the NCPP and largely determined the participants and the subsequent structure. In the middle of 1976, this group decided to tackle most of the entire range of coal-policy issues. The founding group decided to conduct negotiations on strip-mining and land use, air pollution, coal transportation, coal conversion (regulations mandating the conversion of oil- and gas-burning facilities to coal), and electric utility policy, including electricity pricing. The basic idea was to apply the Rule of Reason to coal policy, though there was no inherent necessity to negotiate such a wide range of issues. An experiment might have been restricted to strip-mining policy, for example.

Defining the NCPP as negotiating a wide scope of issues was a risky decision, designed to "put the Coal Policy Project on the map," to make it stand out against the background of other dispute resolution experiments and mediation attempts. Success would make the NCPP seem quite important, and indeed its report, *Where We Agree*, was front-page news in the *New York Times*.[20] A wide scope had the effect of enhancing fund-raising, because the experiment seemed more impressive, and a greater number of corporations and government agencies were interested in donating as the NCPP took on air pollution, for instance.

Opting for the wide scope of policy also increased the chance that the NCPP would reach significant agreements in at least one of the areas of coal policy. On the other hand, opting for the broad scope of issues might have increased the possibility of disagreement, as the NCPP took on more and more issues. Although the founders were optimistic about their endeavor, they still could not be sure that their Rule of Reason idea would work. An intense fight in one area of coal policy might have spilled over into other areas, thereby disrupting the work of the NCPP as a whole. Further, an intense fight might have attracted considerable publicity, more perhaps than any agreements, thereby seriously undercutting the goal of demonstrating possibilities for conflict resolution. As it turned out, such an intense conflict never developed, thanks largely to the atmosphere of agreement and accommodation that built up throughout the NCPP (see Chapter 5). The business side and the environmentalist side fundamentally disagreed about federal policy toward leasing coal-mining rights on federal land, but this basic disagreement did not spill over into other areas of negotiations.

Although Decker, Moss, Whiting, and Murray made a somewhat risky decision by taking on a wide range of coal issues, they did avoid the major area of miners' health and safety and at first partially avoided the coal-leasing issue. As noted above, the leaders factored out the labor-related issues of safety and miners' health because such issues demanded union participation, and it was thought that inviting, say, one member from each of the three factions of the UMW during a union election would result in a stalemate. Decker et al. were probably correct, but while their decision simplified the NCPP's task, it also restricted the project's subsequent claim to legitimacy as being representative of the major groups of the coal issue network.

The four founders of the NCPP, none of whom had long experience in the coal area in its entirety, did not quite appreciate the difficulty of coal-leasing issues and therefore did not give them due emphasis in setting up the NCPP's structure. During the second phase of the NCPP's activity, after the publication of *Where We Agree*, a special subcommittee was established to come to grips with coal-leasing issues.

With the exception of labor issues and the partial exception of federal leasing, then, the founders went after the entire range of coal-related issues. Strip-mining and land-use regulations deal with problems basic to the coal industry and were particular subjects of controversy before the passage of the SMCRA in August 1977. The economics of coal are closely intertwined with the nature of air pollution regulation, since burning coal is a major source of such pollutants as sulfur dioxide, sulfates, particulates, nitrous oxides, polycyclic hydrocarbons, which may be linked with cancer, and various trace metals such as arsenic and cadmium. Coal pollution is related to such diseases as lung cancer, pneumonia, emphysema, and asthma. Sulfates and nitrous oxides are thought by most scientists to be the basic source of acid rain. Atmospheric inversion has trapped coal pollution in the air over cities, causing 4,000 deaths in London in 1952 and 20 deaths in Donora, Pennsylvania, in 1948.[21] Clearly, air pollution regulation must be concerned with the burning of coal.

About 70 percent of the coal consumed in the United States is burned to generate electricity (usually by heating water to steam to turn turbine generators). Coal policy is therefore closely related to government regulation of electric utilities. Consequently, the NCPP became involved in the area of the pricing of electricity by publicly regulated utilities.

The political economy of coal also overlaps the political economy of transportation because of the heavy, bulky nature of coal. Much of the final price of coal derives from transportation costs, generally railroad transportation. Of course, the proportion of transportation cost varies with the distance from the coal fields to the generating facility. Coal from Wyoming is much cheaper in Wyoming than it is in San Antonio or Houston, for instance.

The NCPP also grappled with issues more prominent in the 1970s than today—the issue of mandating a switch from the use of oil and natural gas to coal and various questions concerning coal conservation.

The scope of issues determined the basic structure of the NCPP—a plenum coordinating the work of several committees. The founders thus set up "task forces" for mining, air pollution, energy pricing, transportation, and "fuel utilization and conservation." The complex nature of the several issue areas of coal policy meant sixty people could not, working as a whole, deal with all the issues within a year. Expertise in one area does not readily transfer to another area—strip mining is a matter of geology, air pollution a matter of physical chemistry, biochemistry, and industrial engineering. Government policy concerning air pollution is so complex and confused that few people really understand it. The broad scope of policies selected for the NCPP made the committee structure almost inevitable.

Plenary meetings were chaired by Father Francis X. Quinn. From January 1977 to February 1978, the plenary group met five times for two days each, or ten days in all. The plenary group consisted of twenty people: the business and environmentalist chairs and vice-chairs from each of the five committees. In addition, Decker, Moss, Whiting, and Quinn were members of the plenary body. Endorsement of a proposal required the agreement of a substantial majority on both the environmental and the business side, so the presence of twelve business executives to eleven environmentalists did not matter. The Georgetown CSIS staff defined its role as one of neutrality, so Francis Murray did not serve on the plenary body.

The fact that there were just two "sides" or groups of participants certainly facilitated the negotiation process within the NCPP. There might have been a union side; there also might have been a consumer side, or at least a few consumer representatives on the environmentalist side. In fact, the founders of the NCPP claim that two or three representatives of consumer public-interest groups were contacted, but after some discussion, they decided not to participate. And Decker and the others did not persist in seeking consumer representatives. Whether a negotiating conference like the NCPP should include representatives of consumer groups to be legitimate varies with one's perspective. Many do not consider consumer group staffers to represent the public interest. In any case, it is especially difficult to get consumer group staffers to participate in mediation negotiations like the NCPP for several reasons. Consumer lobbies are small and have few resources. On the other hand, their task is very great—to challenge the health, safety, and pricing flaws of the capitalistic system. Thus, it is not cost-effective for consumer lobbies to spend their limited resources in negotiations. The most efficient way to repair the faults of capitalism is to aggressively seek out, challenge, publicize the flaws in the production of consumer products. Consumer lobbyists are thus the investigative reporters of the lobbying world; their task is not to mediate conflicts, but to initiate them.

Moreover, consumer groups typically do not have a large staff. In 1976, Ralph Nader's conglomerate of about twelve small consumer organizations had a staff roughly equivalent to that of the Sierra Club and a lower total budget—about two million dollars.[22] Consumers Union, technically not a lobby, at that time had only two or three professionals in its Washington office. Given the task and size of consumer groups, it would not make much sense for them to participate in a mediation conference.

One other voice was barred from the NCPP—government. The founders of the NCPP specifically decided not to invite members of Congress, or for that matter, any person holding a government position (including state and local government) or any current elected officeholder to participate in the NCPP negotiations. Of course, inviting civil servants and legislators to participate would have complicated the structure of the NCPP. Representatives of government would necessarily have constituted a third side; virtually none of them would have wanted to serve as a member of the business or environmentalist side, for civil servants and legislators are supposed to make and implement policy for the good of all.

Civil servants in particular would have found it difficult to participate in the NCPP. By February 1977, the NCPP clearly advertised that "the participants in the project took part as individuals. Although they were selected in part because of their leadership roles in environmental and industrial organizations, they do not purport to speak either for their organizations or for the environmental and industrial communities at large."[23] However, civil servants, representatives of the Environmental Protection Agency or the Department of the Interior, for instance, would have an awkward time participating as individuals, since they were also members of the governmental hierarchy that would potentially issue and enforce regulations discussed by the NCPP's negotiators. Ratification of the NCPP's agreements by participant government officials might have seemed to commit government agencies to initiate policies endorsed by the NCPP. Accordingly, superiors of hypothetical governmental participants would instruct them to clear agreements with agency officials, and it is unlikely that such officials would give such approval without clearance from the agency head and from the Office of Management and Budget, which obviously might block agreement or take years to give their assent. There are thus strong reasons for excluding government officials from NCPP-type negotiations, although in similar types of alternative dispute resolution, government officials can play a critical role (see Chapter 8).

One might have expected legislators interested in coal policy to participate. Such members of Congress in 1976 as John Dingell (D-Mich.), Morris Udall (D-Ariz.), or Paul Rogers (D-Fla.) come to mind. Legislators might have contributed a great deal, but they had to be excluded to maintain the nonpartisan image of the NCPP, a major goal of the project's leadership. If the NCPP were viewed by Washington decision makers as specifically Democratic or Republican, liberal or conservative, its recommendations would become just another part of the partisan battle. In fact, most of the NCPP's participants lacked politically partisan images. True, the majority had some public relations or lobbying experience; political participation in interest politics, as opposed to party politics, was helpful to the NCPP, which purported to bring together interests and mediate conflict among them.

Beyond these factors, the original conception of the NCPP precluded government participation. In 1976 the founders believed that if a grand coalition of environmentalists and coal-related business leaders could agree on a broad platform of recommendations, the legitimacy of such an agreement would be so great, and its inherent logic so powerful, that Congress and executive branch officials would rapidly adopt the NCPP's proposals. Accordingly, what little need there was for government officials to participate was greatly outweighed by the additional difficulties it would cause, in the view of Decker, Moss, et al. Furthermore, business participants recognized environmentalist lobbies as equal and legitimate participants in the private direction of policy. A theme of the NCPP was that an industry-environmentalist coalition "knew best," and was better able to initiate consensus policies in the public interest than were government "bureaucrats."

Larry Moss and Jerry Decker both had considerable influence in picking the persons who participated in the NCPP. In inspecting the chart of participants (Appendix A), one notices that three leading participants were contacted through the Sierra Club, of which Moss had been president in 1973–1974. An initial cochairman of the mining task force was J. Michael McCloskey, executive director of the Sierra Club. Gregory Thomas, vice-cochairman of the transportation task force, was then a lobbyist for the club. Bruce J. Terris, cochairman of the air pollution task force, headed his own public interest law firm in Washington, D.C., and had worked with the Sierra Club since the late 1960s, conducting some of their major lawsuits. Other environmentalist groups with an affinity for the NCPP were the Environmental Defense Fund, a group best known for lawsuits regarding the use of hazardous chemicals, and the Environmental Law Institute, a library and communications center concerned with the rapidly changing nature of environmental law.

Decker and Whiting recruited a temporary business steering committee, composed of executives from coal-related industries, to help them recruit executives for the business side of the NCPP. First, we should note that the mining task force had members who were production experts from the three major coal companies: Peabody Coal Company (first), Consolidation Coal (second), and AMAX (third). The group contained the chief executive officer of Peabody, Edwin R. Phelps; Peabody's chief lobbyist, Harrison Loesch, who was also the lead lobbyist for the coal industry on strip-mining legislation; a former chairman of Consolidation Coal (a branch of Continental Oil), John Corcoran, who was perceived to be an industry leader; and representatives from AMAX, a minerals conglomerate controlled by Standard Oil of California.

Second, although lawyers were placed on the transportation task force, Decker urged that engineers and economists staff the air pollution, fuel utilization and conservation, and energy pricing task forces. Significantly, three of the four chief founders of the NCPP—Decker, Moss, and Whiting—were trained in science and engineering. (Frank Murray had a graduate degree in business administration.) One strategy informing the choice of participants was the hope that engineers and technicians on the business side would be able to communicate about technical issues with such scientifically trained environmentalists as Robert R. Curry, a geology professor at the University of Montana, and J. Russell Boulding, another geologist, who succeeded Curry as environmentalist leader of the mining task force. In accordance with Wessel's Rule of Reason, this approach emphasized the value of factual communication over strategic maneuvering.

A third and final noteworthy aspect of the recruiting of businesspeople was a problem arising from conflicts among corporate interests competing for the lucrative business of transporting coal. Most coal is transported by railroad, although some is transported by river barge, and the fledgling coal slurry pipeline industry is in sharp conflict with the rail industry over the question of pipeline acquisition of eminent domain rights to cross a railroad's right-of-way. The rail industry has successfully lobbied Congress to block a federal law granting slurry pipelines a right of eminent domain.[24] Further, during the time of the NCPP, the rail industry and many others were seeking a major increase in river tolls for barges, on the grounds that the barge industry was federally subsidized because river dams and shipping channels were constructed at federal expense. (Owing to its bulky nature, coal is trucked for only short distances, such as from mine to railroad.)

The founders of the NCPP felt that if executives from railroads, the barge industry, and slurry pipelines were present on the business side of the transportation task force, corporate viewpoints would be vanquished by a united environmentalist side, leading to a false solution with little appeal to industry. The NCPP's leaders, accordingly, decided to put no representatives of transportation industries on the task force. The result, however, was a disappointingly vague agreement on a general platform of government deregulation of transport that lacked the specific punch of NCPP reports on strip mining or air pollution (see Chapters 5 and 6).

In terms of individuals' incentives to participate in the NCPP, by fall of 1976 the project seemed likely to be a significant (and hence attractive) activity because (1) the initial leadership seemed competent, (2) Georgetown CSIS staffing and financial support from the Ford and Rockefeller Foundations provided organizational resources, (3) the test meeting in July 1976 boded well for the group, and (4) the Rule of Reason process provided a specific procedure. A number of other motives were found among participants: an agreement with the idealistic message of the NCPP that alternative modes of conflict resolution were needed to stabilize the coal sector and to help develop American energy independence and a desire to "make a contribution" to society, for one. For some members, curiosity was a motive: they wished to find out if the idea could work. Some were also curious to meet their opposition: environmentalists and their business opponents ordinarily do not know one another personally. As a participant from New Mexico Citizens for Clean Air and Water stated in 1979: "The Coal Policy Project is worthwhile just from an educational standpoint. People learn more about one another's motivations and situations. Sometimes the positions which are taken by business seem irrational. But they become reasonable once the positions are explained."[25] Friendship for either Jerry Decker or Larry Moss was another incentive to participate.

Monetary incentives had some importance. Business participants drew their regular salaries during their several weeks of participation. Some businesses probably took a tax deduction for these costs. Environmentalist lobbies, on the other hand, have neither the staff nor financial resources to pay people to engage in negotiation experiments. Accordingly, environmentalists were paid $150 per day in consulting fees for participation. To some environmentalists, the NCPP was a source of extra cash. These consulting fees came from foundation grants, rather than from business contributions, to ensure that the environmentalists did not appear to be co-opted by business money.

Several businesses, whose executives participated in the NCPP, no doubt wished to participate in the NCPP to demonstrate their concern for the public interest. Such businesses probably included the three major coal companies and Dow Chemical. Pacific Gas and Electric, which, like Dow Chemical, had been interested in sponsoring alternative energy production and conservation experiments, provided two participants.

Decker and Moss were aggressive recruiters, and after the initial organizational spadework had been done, there was a considerable incentive for business executives and environmentalists to join the NCPP process. Nonetheless, there were some very significant refusals. Moss had hoped to recruit Louise Dunlap, head of the Environmental Policy Center (EPC), the leading environmental lobbyist on strip-mining issues. But Dunlap unequivocally refused to join the NCPP. The whole effort was not to her taste. Dunlap saw herself on the cutting edge, an initiator of strong environmentalist positions, and a fighter in her position as lobbyist to get environmental legislation enacted. It seemed to her contradictory to engage in a mediating effort to compromise legislation that she had fought for in Congress, while some of her supporters in the field were receiving anonymous threats of bodily harm.[26] Dunlap not only refused to participate, but saw the NCPP as positively harmful to progress in the environment, and accordingly set out to undercut the NCPP, with considerable success (see Chapter 7). Moss eventually recruited Michael McCloskey to head the mining task force, but Dunlap finally got the Sierra Club Board to withdraw McCloskey's participation (see Chapter 7).

Another defeat for Larry Moss was Richard Ayres's refusal to participate. Ayres, affiliated with the Natural Resources Defense Council, was generally recognized as the leading environmentalist litigator on air pollution issues and was universally respected for his expertise on the subject. Ayres did participate in two early NCPP meetings, devoting a few hours of his time, but then withdrew, stating that he did not expect the outcome of the project to be worth the time expended in participation. In addition, he seemed to suspect the motives of the NCPP organizers. As a litigator on issues involving stakes of billions of dollars, Ayres and the NRDC assumed major responsibilities; accordingly, his point of view that the NCPP was "small fry" has considerable justification. At that time Ayres was involved in negotiating an out-of-court settlement with the Tennessee Valley Authority about the installation of scrubbers in several coal-burning generating plants, a question involving the possible expenditure of nearly a billion dollars. Moss eventually recruited Bruce Terris, perhaps the nation's number-two environmentalist litigator on air pollution issues, to head the NCPP's air pollution task force.

Given Dunlap and Ayres's refusal to participate in the NCPP, one might ask why significant environmentalists did agree to work on the project. Besides the five motives discussed above—idealism, belief in the project's workability, profit, curiosity, and friendship for Larry Moss—two additional motives deserve mention. In retrospect, the project's director Frank Murray commented:

> Some of the environmentalists participated because they wanted a place at the table. The environmental lobbies were new in 1976. Not everyone accepted them as players in the national game. In the Coal Policy Project the environmentalists were treated as equal to the corporate executives. This gave the environmentalists a sense of their own legitimacy. The signal was that they were part of the political leadership equal to any other group.[27]

Some environmentalists, then, in Murray's view, were pleased that the project indicated that the environmental movement had arrived as an important political force. Murray continued:

> By 1976 environmental leaders realized that sometimes the tables were turned. In the beginning the environmentalists were out of power and used the strategy of delay and delay to stop business from doing things opposed by environmentalists. But by then the environmentalists were gaining power in government. Pro environmental legislation was being passed or about to be passed. Regulations were written under these laws, and now it was the turn of business to go to the courts and use the strategy of delay to slow down the enforcement of environmental laws. Business felt justified in doing this because they thought that environmental rules were often unrealistic.[28]

For instance, environmentalists knew they had the power to pass a bill to control strip-mining, but they also knew that the enforcement of such a law faced difficulty because of the litigation and lobbying strategies of business. This provided an incentive for environmentalists to negotiate with business some mutually satisfactory arrangements in the implementation phase of environmental legislation.

Appendix A lists the participants in the first phase of the NCPP, the period lasting from January 1977 until February 1978, when *Where We Agree* was issued to the public. A second series of meetings was held from January 1979 until March 1980; the studious reader may consult *The National Coal Policy Project: Final Report* to get a list of second-stage participants. The leadership changed little in the second phase: eighteen of twenty-four members of the second-phase plenary served on the plenary in the first phase. Of the six new plenary members, five had been NCPP participants in the first phase. The number of negotiators was limited to thirty on each side during the first phase, but the total number of participants in the NCPP totaled 119, including alternate negotiators and a panel of twenty industry experts recruited to advise the industrial members of a cogeneration task force in the second phase, and the twenty-one new negotiators participating in the second phase. Appendix B lists meetings of the NCPP for the first phase. Active participants during 1977 needed to put considerable time into the project; particularly the twenty-four members of the plenary group, who also served as a cochair or vice-chair of one of the task forces. Members of the mining task force had to do considerable traveling in order to learn about the variety of mining problems in different geographical circumstances.

FINANCES

Complete records of the financing of the NCPP are no longer available, but I have obtained copies of some of the financial records. No "big scoop" here, no inside story, but there are a number of interesting details. The project finally spent about $1.4 million. Of this amount, businesses contributed $792,600, or 60.3 percent; foundations contributed about $350,000, or 23.6 percent; and government agencies gave $225,000, or 16.1 percent.

Of the eighty-six corporations that contributed, most were electric utilities or chemical, steel, coal mining, and oil industries. This is no surprise; as coal-related industries, they were approached by fundraisers Murray, Decker, and Whiting. No railroads contributed. Apparently the fund-raisers avoided the railroads for the somewhat ironic reason that *Where We Agree* was prorailroad. It argued for deregulation of the rail industry and for an increase in river tolls for the barge industry. Representative Doug Walgren, a Pittsburgh Democrat sensitive to the interests of the river barge industry, publicly charged the NCPP with being financed by the railroad industry; a complete lack of railroad contributions enhanced the credibility of *Where We Agree.*[29]

The eighty-six businesses contributed an average of $9,000 each. Contributions were tax deductible and could be cited by corporate public relations departments as an example of a company's responsible concern for the public interest. Apparently no single company gave more than 5 percent of the $792,000 total corporate contributions. Records of business contributions for 1977 and 1979–1980 are available, but records for 1978 are lost. For the three years cited, however, the major corporate contributors were Dow Chemical, $30,000; Dresser Industries, $25,000; Texaco, $20,000; U.S. Steel, $20,000; PPG Industries, $20,000; DuPont, $15,000; General Electric, $15,000. In addition, business picked up the salaries of business participants, although such salary expenses were probably written off by most companies as a charitable contribution. Dow Chemical and Peabody Coal seem to have contributed the most executive time to the NCPP.

As for foundation contributions, $150,000 came from the Andrew W. Mellon Foundation, almost half of the total of $330,000. The William and Flora Hewlett Foundation gave $60,000; the Ford Foundation, $50,000; the Rockefeller Foundation, $50,000; the Sarah Scaife Foundation, $10,000; the Brown Foundation, $10,000. The Mellon Foundation seemed to exercise no particular influence over the operations of the NCPP. The Rockefeller and Ford foundations gave smaller amounts, but exercised influence over the NCPP by their early encouragement of the founders to go ahead with the effort. Theirs was a "seed money" role. The two staff directors, Murray and Curran, saw the Ford and Rockefeller contributions as critical to the NCPP: "Without their assistance it is probable that the project would have been abandoned," they wrote.[30]

About half of the $225,000 from government agencies came from the Department of Energy. The DOE was not observed to have influence over the progress of the NCPP, although that department ordinarily sent several observers to plenary meetings. The first secretary of energy, James Schlesinger, was supportive of the concept of the NCPP, but he did not push for implementation of the NCPP's proposals with his department. A subunit of DOE, however, the Federal Energy Regulatory Commission (FERC), was the one federal agency most receptive to suggestions from the NCPP.

Although records of expenditures are no longer available, a good idea of the pattern of expenditures can be gained from the estimated budget for the calendar year 1979, when the second phase of meetings was held:

Plenary group meetings	$48,000
Task Forces: meetings and field trips	70,000
Task Force technical staff	50,000
Support for environmentalist side	72,600
CSIS staff and administrative support	97,360
Other (reports, communications, miscellaneous)	25,000
Subtotal	$362,960
CSIS overhead (17.4 percent)	63,155
Total	$426,115

In other words, expenditures for plenary and task force meetings involved air fares and hotel bills. Many participants lived far from Washington, and field meetings were rather expensive. Consultants were hired to give advice about the complexities of air pollution policy, the geology of strip-mining, cogeneration engineering, and so forth. Georgetown University's CSIS is located at a convenient, prestigious, and expensive location in the heart of Washington's famous K Street sector. The final report of the NCPP was not published until 1982, and thus the NCPP was responsible for at least half of the salaries of the two managers, the Director, Francis X. Murray, and the Associate Director, J. Charles Curran, for the six-year period of 1976–1982. The project also employed a full-time government liaison, Ralph Nurnberger, for two years, and several administrative and clerical employees. And the long-distance telephone bill was large. Nor must one overlook the donated time of corporate executives. Overall, the NCPP was a rather expensive experiment (consult Appendix C for a list of contributors).

CONCLUSIONS

The original organizers—Decker, Whiting, and Moss—were quite successful in persuading others to back an experimental negotiating body based on the principles of the Rule of Reason. The concept of the NCPP interested the Ford and Rockefeller foundations, as well as the Georgetown Center for Strategic and International Studies, which backed the effort with staff and meeting facilities. The organizers were able to locate about sixty businesspeople and environmentalists of some stature who were pragmatic enough to be interested in participating. Further financing came from small and medium-sized grants from dozens of businesses, the Mellon Foundation and a few other foundations, and the Department of Energy and other governmental agencies.

Some organizational decisions contributed to later difficulties (see Chapter 7). Two environmentalist leaders refused to participate, slightly undercutting the legitimacy of the enterprise. Government officials were not invited to participate because both the business and environmental sides believed governmental officials would detract from successful negotiations. The organizers decided to tackle a wide scope of issues, which sometimes resulted in general, vague platforms about transportation and some other issues. Little attention was given to the issue of how the NCPP's platform, once agreed upon, might be implemented (see Chapter 7).

Most important, however, the successful organizing effort demonstrated that in 1976, talented organizers could find without great difficulty prestigious participants, foundation and corporate financial support, and university staffing for an experimental project in intergroup negotiations to find common policy interests. Interest in similar experiments flourishes in the 1990s (Chapter 8), and the organizers of the NCPP deserve credit for launching a pioneering effort.

5
The Process of Consensus

A SOCIAL SYSTEM FOR AGREEMENT

As noted in Chapter 2, the air pollution, strip-mining, and other coal-related issues in the 1970s were characterized by complex, disputatious, drawn-out conflicts, which induced some businesspeople and some environmentalists to desire a better way of making public policy. The Rule of Reason provided a set of effective organizing principles to establish a negotiating body among contending environmental and business groups. Talented organizers were able to put together the participants, money, and staff needed to seek common interests among the normally opposed interest groups. In addition, the emergent leadership of the NCPP invented a system of norms, supplemental to Wessel's norms, and developed an organizational structure that greatly enhanced the process of reaching agreement.

The creators of the NCPP—Decker, Moss, Whiting, Murray, and Quinn—established a social system for producing agreements between persons normally in conflict. We are beginning to see how this system worked. First, an attractive set of goals was posited: establish a national policy mediation conference to deal with coal-related issues that would point the way to energy independence and demonstrate a new and better way of resolving social conflicts. Decker discovered a text that set forth the basic norms for behavior in the national policy conference: (1) respect the people and groups on the opposing side as representing with integrity a significant set of interests, and (2) be forthcoming with information about tough, technical issues rather than treating information strategically. Consequently, behavior according to the norms fulfilled one of the goals of the group: to demonstrate the possibility of resolving energy/environmental conflicts outside the context of adversarial institutions.

If the NCPP represented a joining of these goals with the Rule of Reason norms, it was indispensable to the process of consensus building that the leadership group be very effective in creating the new institution. After the second, successful test meeting, the behavior of the five leaders was critical if Wessel's norms were to help meet the goals of the NCPP. These five people believed in the possibilities of the new social process, convinced others it would work, themselves served as models of Rule of Reason goal seekers, and trained various task force leaders to follow Wessel's negotiation rules and to convince others to abide by them.

Again, it seems that Decker especially played a key role. It is not surprising that he was able to persuade another leading executive of Dow, Mac Whiting, to help in establishing the NCPP, and it was certainly fortunate that he was able to link up with Moss. Decker was a natural choice for the government advisory boards to the Federal Energy Agency (a predecessor of the Department of Energy), and thus he was bound to meet several leading environmentalists, including Moss. Although most corporate executives never get to know an environmentalist leader firsthand, Decker was in a different position. Yet perhaps there was also an element of luck in finding Moss, as he was interested enough in the NCPP idea to spend much time and energy to help establish the experiment.

Larry Moss had served as president of the Sierra Club in 1973–1974, and he had helped initiate one of the most significant environmental lawsuits in history (*Sierra Club v. Ruckelshaus*) to block the deterioration of air quality in areas cleaner than the legal standard.[1] He seemed to differ somewhat from most environmentalist lobbyists in Washington as he was neither associated with the counterculture of protest nor a lawyer. Instead, Moss had acquired degrees in chemical engineering and nuclear engineering, and he had been appointed a White House Fellow during the Johnson administration, an honor usually taken to mean that the recipient is likely to be a future top-level leader. As an engineer, Moss liked and understood technology. He was also an enthusiastic aviator and a proponent of emphasizing an economic approach to environmental issues. For instance, he was an aggressive proponent of the emissions tax approach to pollution control, one that would permit the business decision maker to decide how much pollution to release and then pay a federal tax for each unit of pollution. An environmentalist and an engineer, Moss was also interested in politics. In one task force meeting I observed, he seemed particularly motivated to try to draw up compromise proposals, involving complex, technical issues, that would be pleasing to both sides.

The combination of Decker, Whiting, and Moss was likely to appeal to one or two foundations and/or a research institute, and indeed it did. Although Frank Murray, the NCPP's administrative director, would demur at being included in a leadership group that set up the NCPP, his fund-raising activities, advice to Decker and Moss, and general enthusiasm for the project qualify him as one of the leaders. Finally, as labor mediator, Father Francis X. Quinn was able to see the value of the Rule of Reason as a conciliating device and the virtue of working to implement it.

The consensual social system of the NCPP was enhanced by a selective recruitment of participants. Obviously, the leadership group screened out particularly abrasive or confrontational personalities where possible. Moreover, such persons were unlikely to be interested in participating in the NCPP anyway, thereby reinforcing the process of selective recruitment. Consumerists, whose Herculean role of challenging the entire capitalist system promotes aggressive, confrontational behavior, chose not to participate also because their small organizations could not spare personnel.[2] Union leaders, fighting among themselves, were not asked to participate. Conversely, businessmen and environmentalists who chose to participate were likely to be especially interested in conciliatory approaches to conflict and particularly willing to accept Wessel's rules. They were not interested in agreement just for its own sake, however, as virtually all of them had shown some previous commitment to either environmentalist or business causes.

As part of the selective recruitment process, certain issues were factored out of the NCPP's discussion process. The NCPP did not try to deal with complex, controversial labor issues. Consumerists would surely have objected to increasing consumer prices in order to enhance free markets but this conflict was factored out of the process by selective recruitment. Finally, in one major area, the phase-one negotiators agreed to disagree over an issue that baffled the negotiators—the development of a rational policy toward federal government leasing of its coal lands in the West.

The existence of an effective leadership circle prior to the first plenary meeting catalyzed the process of recruitment. Decker, Moss, et al., were able to convince participants that the Rule of Reason approach had a good chance of working, contrary to the initial reaction of the typical business executive or environmental lobbyist. Through a sort of salesmanship, Decker et al. were able to dispel natural doubts about the Rule of Reason, convincing participants to give it a serious try, thereby getting the consensual process off to a good start, which in turn encouraged participants further to believe that the process of agreement might work.

Supplementing the Rule of Reason, Decker, Moss, et al. promulgated three other sets of norms: the norms of economics, the norms of science, and the norms of voluntarism. The norms of economics are described further in the next chapter, but the essential point is that Moss, especially, pushed for the recognition of generally accepted views within the field of economics as bases for agreement. During the Carter years, both Democrats and Republicans, the president and his conservative opponents, occasionally demonstrated an ability to agree on principles of deregulation.[3] Conservatives felt that it would be a good thing to shrink the size of government and to promote competition. Liberals felt that special-interest combinations of government and business were exploiting the consumer and thus sometimes deregulation would be a good thing. Moss promoted similar attitudes to enhance consensus within the NCPP. Not only deregulation, but emissions taxes, cogeneration, user siting of power plants, and marginal-cost pricing for electricity could be promoted as the "reasonable" and "technically accurate" joint solutions for controversial issues. The point of view of the economist sometimes attracted business more, and sometimes the environmentalists (e.g., it bolstered the arguments for cogeneration). This topic is developed in Chapter 6.

The norms of science also promoted agreement. They emphasized the need to understand the factual background of some of the issues, and the belief that a joint agreement on most of the "facts" might be possible and facilitate agreement on policy issues. Decker, Whiting, and Moss had degrees in engineering or science; Murray had served as an executive in a science-related federal agency. Due to the nature of the leadership and to some selective recruitment, the NCPP may have had a greater tendency to emphasize technical details than the typical, Washington-based negotiating group.

The norms of science were more dominant in government during the Progressive reform era (1900–1917) than today; then it was generally assumed that scientific administrators could discover the facts of a case and define a correct course of action above the divisions of politics.[4] The NCPP to some degree shared this Progressive belief; in turn-of-the-century terms, the NCPP hazarded the idea that facts might establish a consensus on a correct course of action in the administration of strip-mining, air pollution, and electricity pricing policies. Today, the norm of scientific administration seems naive to the political scientist. Nevertheless, it should not be ruled out as totally inapplicable. In the case of the NCPP, agreement on the "fact" that Northern Plains strip-mined coal had a low heat content per pound meant both business and environmentalists could agree on the "fact" that less Northern Plains coal would be developed immediately than many experts had once predicted, and this prospect made it easier to agree on a platform for regulating such development.[5]

The norms of science overlapped the norms of economics, because Moss and most participants actually viewed economics as a kind of science, which set forth concepts and facts that rational persons could agree on. More important, the Rule of Reason overlaps the norms of science because the Rule of Reason attempts to improve communication among a community of professional truth seekers, an endeavor resembling rather the norms of science than the adversarial legal process. Scientists do not ordinarily manipulate their data to convince some third party that one or the other experiment is correct. Intentionally omitting data to enhance a particular interpretation violates scientific norms, and scientists who are caught in such behavior suffer a loss of reputation. The Rule of Reason exhorts lawyers to behave more like scientists, or at least more like ideal scientists. Wessel believed that legal truth seeking is enhanced when participants make as much data as possible known to the entire community as early as possible. Wessel argued that discussions about the meaning of such data should be held early in the process, to ensure that the litigants and the public arrive at a satisfactory joint approximation of justice. Just as scientists respect one another's integrity as truth seekers, so litigants should respect one another in a truth seeking process. The Rule of Reason is likely to appeal intrinsically to those trained in science and engineering. One can see how the Rule of Reason, the norms of science, and the norms of economics would interact in a powerful, reinforcing cultural system. We begin to understand why it was possible for the NCPP to succeed so impressively in finding agreements.

A third norm for consensus building was the norm of voluntarism, a belief generally shared by the business side and vigorously advocated by Larry Moss and some of the environmentalists such as J. Michael McCloskey, that it was better to make decisions in the NCPP, a voluntary, private body, than to make such decisions in government, particularly in the bureaucracy. One unifying theme, stated by Moss, was that the NCPP's negotiators needed to reach agreements or else solutions would be promulgated by federal bureaucrats, and both business and environmentalists would lose. This was a form of Paretian reasoning: essentially it argued that most compromises of the NCPP were Pareto optimal decisions, as both sides were better off when compared to the alternative, bureaucratic fiat. Accordingly, the norm of voluntarism had depth; it was not only a belief that encouraged cooperation and high morale—"our decisions are better than bureaucratic decisions"—but it also appealed to the rational, self-interested behavior of both sides—"we must compromise with the other side or else face a government-imposed solution, which is sure to be worse than the compromise." Of course the norm of voluntarism assumed that the NCPP's decisions would become public policy, which turned out not to be the case.

Note that voluntarism overlapped the norm of economics and science. Economists tend to argue that decisions made by private markets, or by marketlike behavior, are preferable to decisions made by government through administrative orders. Most of the NCPP's participants accepted this and accordingly agreed on such items as deregulation in transportation, emissions taxes, and so forth. Such ideas from economics give more decision-making latitude to nongovernment actors, reinforcing the idea of voluntarism. Similarly, the norms of science, as interpreted by the NCPP'a participants, tend to favor truth seeking among interacting individuals freed from central direction or authoritative hierarchies. Instead of the economists' market model of economic exchange, science prefers the decentralized interaction of individuals communicating theory and information in search of truth. Both ideas reinforce the norm of voluntarism.

Voluntarism is further discussed in Chapter 6, but we should note here that the NCPP's participants believed if they could reach agreements from the perspective of varied environmentalist and business backgrounds and interests, the attractiveness of such joint compromises would virtually compel adoption and implementation by Congress and federal regulation writers. During 1977, most of the NCPP's leaders saw themselves as participating in the construction of a grand coalition of business and environmentalist groups that would compel agreement from government. The NCPP's leaders were first-stage interest-group theorists who believed that the important variable in determining policy is group pressure—hence, if major groups agree, that determines policy—and therefore downplayed the independent role of agencies of the state. And in the terms of Theodore Lowi's *The End of Liberalism*, they were ideological pluralists, who preferred that administrative decisions be determined by the balance of power of pressure groups, rather than by authoritative administration of public law.[6] If the NCPP's participants had been able to win the backing of their organizations, perhaps a more successful grand coalition of groups would have formed and voluntarism and group theory would have described the new public practice. Although this did not happen, joint belief in the norm of voluntarism during the negotiating process contributed to group solidarity and the search for agreement.

At this point, another factor becomes important. Members of the NCPP, many of whom had never met before, came to like one another. In many instances, business executives and environmentalists became friends. The small working groups and the apparent success of the overall organization produced an atmosphere of friendliness. This of course was another important cause of reaching agreement. Friendship is normal in such circumstances, but perhaps especially so in the NCPP, where the goals and norms of the group emphasized the importance of respecting one's opponent in a joint process of truth seeking to reach agreement. The whole social system of the NCPP seems designed to make friends out of the participants.

In particular, members of the mining task force became good friends. This goodwill helped greatly in elaborating a long, detailed, patiently crafted report that dealt with some very controversial aspects of strip-mining policy. The leaders of the mining group were scheduled to spend twenty-four days meeting during the thirteen months of the NCPP's first phase. Among the more interesting meetings were group field trips to learn about the various conditions of strip mining in different parts of the country. The experience of the mining task force in journeying to the somewhat remote mining regions of the Northern Plains was an example of how persons of different backgrounds can develop group solidarity through participating in a common task while isolated from their peers. The story of how the mining task force wandered through the rain in Billings, Montana, searching for the El Dorado of a Japanese restaurant came to symbolize the positive, cooperative aspects of the NCPP.

The basic structure of the NCPP tended to produce agreement. Of course it would have been unlikely for the thirty conferees on each side to work out so many agreements if they had met as only a plenary body. Much of the subject matter related to coal policy is quite technical, and specialized committees were necessary. The leadership group also corrected the imbalance of numbers of the initial, failed test meeting. Meeting in committees, having five to seven persons on each side, was conducive to agreement. Such persons could become specialists and develop group solidarity within their own small group.

What is less obvious is that a norm of committee autonomy developed. Leaders of one committee did not seek to influence the development of reports from another committee. During the plenary meetings in 1977 when reports from the task forces were submitted, the assembled body of committee chairs and vice-chairs did not seek to make major modifications in the reports from the various committees. The task forces were viewed as separate and autonomous working bodies, and outsiders did not interfere with these workings nor seek major changes in the agreements drawn up among the negotiators in committee. In turn, task force chairs sought to achieve as much unity as possible in their reports and generally succeeded. There are, however, a few passages in *Where We Agree* noting that individual members dissented from a particular section of a report, or that environmentalists or business executives could not completely accept a point. This unity within the committees contributed to the adherence to the norm of autonomy of committees, because the less division within a committee, the less motivation for other NCPP leaders to become involved in a committee, particularly since much of a committee's report was difficult to criticize by outsiders lacking sufficient technical knowledge.

Each task force submitted a single, unanimous report (with specific disagreements noted). The plenary group then adopted the five task force reports, plus the ad hoc committee (emissions tax proposal) report. The entire body of *Where We Agree* was thus ratified by the twenty-three group leaders, not by all sixty participants. Everything was adopted unanimously, except for the transportation report. Three leaders of the mining task force—Corcoran, McCloskey, and Phelps—filed a minority report, objecting that the transportation statement "contains recommendations regarding broad economic issues" going beyond "concerns with the environment, coal production, and the efficient use of coal," the proper task of the NCPP.[7] This minority report had to be printed in *Where We Agree: Summary and Synthesis*, and thus gained significant circulation.

Yet, the general tendency to establish a division of labor within a group and not to seriously question the results of that division of labor obviously contributed to overall unity and to the possibility of releasing the massive, detailed statement of recommendations found in *Where We Agree*. If several of the NCPP's leaders had tried to alter the process of discussion within the discrete committees, or to launch major efforts to rewrite the reports of these committees, consensus would have been broken, disagreements would have increased, and the project might well have failed.

We are now in a position to understand why the NCPP was successful in reaching agreement among its negotiators. The factors mentioned in this chapter are summarized in Figure 5.1.

COMPARISON TO THE HOUSE APPROPRIATIONS COMMITTEE (HAC)

Some readers have noticed that the consensual process of the NCPP has much in common with the agreement process within the House Appropriations Committee (HAC) of the U.S. House of Representatives in the 1950s and 1960s, as analyzed in the classic study by Richard Fenno.[8] The similarity in consensus building in two rather different bodies such as the NCPP and the HAC implies the existence of general principles that could be applied to the study of other committees and groups or to the creation of such groups when they do not yet exist.

Briefly, Professor Fenno was impressed by the effectiveness of the HAC in performing its obviously difficult task. Recommendations by the full HAC are almost always accepted without significant change by the full House. Appropriations is at the heart of the political process, and accordingly, the HAC has considerable political power, because it determines House appropriations figures (which then had to be bargained with the Senate). Many observers regarded the HAC as an effective institution during the 1950s and 1960s, because it played an important role in keeping down the expenditures of the federal government, which are subject

Figure 5.1. Summary of the Consensual Process

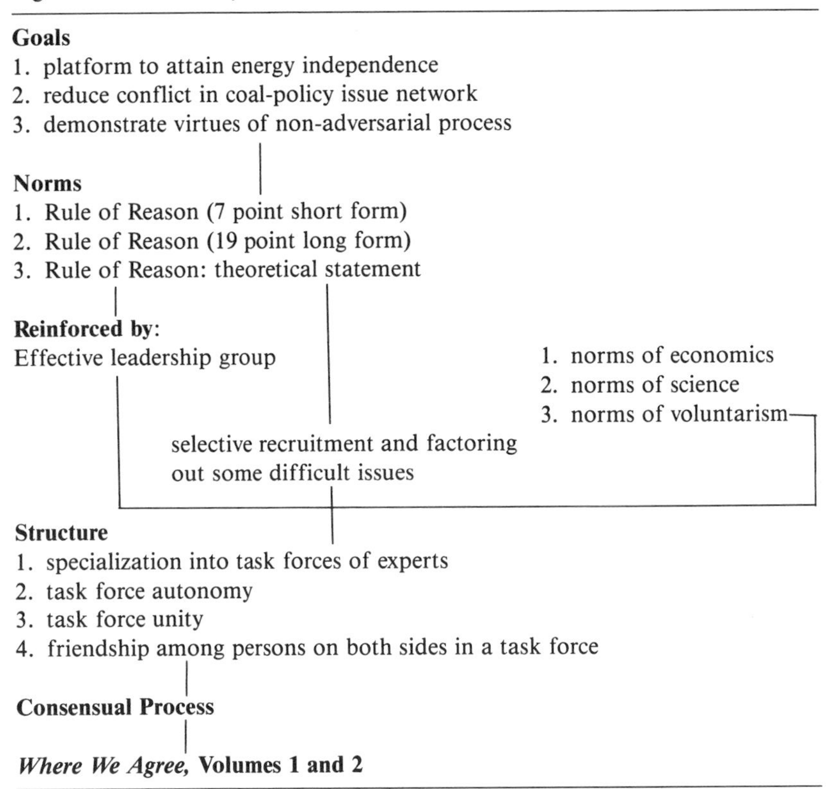

to constant pressures for increases because of the particular interests of groups, agencies, and reelection-oriented legislators, often working together to increase expenditures in areas of mutual interest. By the late 1970s, there was some reason to believe that the consensual process described by Fenno had decayed; perhaps one factor in the increasing instability of the federal budgetary process.[9]

Fenno found that the HAC had clear goals which were articulated with norms of behavior for members of Congress serving on the committee. The chief goal of the HAC was "to guard the Treasury" against what was seen to be a horde of special-interest pleaders. Another goal was to preserve the power of the HAC within the appropriations process, a goal that would be undercut by divisions within the HAC. These goals were connected with a number of general norms of behavior: congressmen were to work hard and conscientiously on the committee; congressmen were to observe seniority on the committee, not only in accepting seniority as a rule for appointment to subcommittee and committee leadership positions, but also in the sense of an apprenticeship rule for new members of the committee. New members had to ask questions last and were not supposed to engage in long speeches, grandstanding, or dramatic behavior in hearings. The goals and norms of the HAC were reinforced by an effective leadership group: the chair and subcommittee chairs, and the ranking minority members of the committee and subcommittees.

Agreement on the HAC was fostered by selective recruitment. The leaders of the HAC and their friends in House leadership positions refused to place new members of Congress on the committee; instead, new members had to demonstrate their seriousness and likelihood of adherence to the HAC norms by serving two or three terms before being appointed. Members who were ideological, abrasive, uncooperative, lazy, or publicity seekers were not appointed to the committee. Nor were members who were subject to close elections in their district. Electoral competition might undermine a member's loyalty to the HAC and its norms by encouraging a member to seek publicity and to press aggressively for special favors for his district, thereby disrupting the spirit of unity and seniority. (Mind you, HAC members got plenty of porkbarrel from committee service, but they were supposed to wait their turn.)

Other norms of behavior on the HAC were linked to its structure as a working body. These norms often resembled those of the NCPP. The HAC had occasional full committee meetings dominated by the subcommittee chairs, similar to the NCPP's occasional plenary meetings composed of its task force chairs. Most of the work of the HAC was done in its thirteen or so subcommittees, one for each of the separate bills of the budgetary process. The total size of the HAC, about fifty-five, was close to the sixty members of the NCPP. Both broke up into subcommittees of about seven members. (HAC members served on two subcommittees; NCPP members on only one.) Fenno described behavior within a subcommittee as following the "norm of minimal partisanship" because congressmen within the group tended to be friends and to defer to each other, regardless of party. Similarly, the NCPP was organized as a two-party system, having a business and an environmentalist side, and similarly, the two parties followed a behavioral norm of minimal partisanship. In the NCPP, this meant that people on one side were expected to consider seriously factual presentations and policy proposals made by persons on the other side. Both sides were expected to present serious compromise proposals, not incorporating strategic elements intended to benefit just one side.

In both the HAC and the NCPP, the result of these norms was subcommittee unity, which itself became a norm as members tried to reach unanimous agreement on a report. Given unanimity, the full committee or plenary was hard-pressed to question the report of its subunit, especially as such reports involved complex, and often technical, information. The HAC also followed the norm of subcommittee autonomy; members not serving on a subcommittee did not attend its hearings and seldom tried to influence its budgetary requests. (The chairman and ranking minority leader had this prerogative, however, just as cochairs Decker and Moss had the prerogative of sitting in on task force meetings.)

The HAC also factored out some complex issues that would have disrupted the process of agreement. As shown by Aaron Wildavsky, this process was achieved through "incrementalism"; the HAC took for granted the previous, or "base" budget figure, and concentrated its analysis and discussion on the yearly increment requested by an agency.[10]

The consensual process had a different effect in the HAC case because of its standing as an authoritative institution of government. In unified fashion, the HAC presented its budgetary proposals to the full House, which in most cases made little change in the HAC figure. The HAC budget figures thus became a base point in the budgetary process, in which the Senate ordinarily proposed a different figure, and the president retained the threat of veto of an appropriations bill, especially if it were too high. The NCPP, on the other hand, was not linked to authoritative institutions, so its proposals were largely ignored.

Of course there are limits to the comparison of the NCPP and the HAC. Even in the 1950s, we might expect greater conflict within the HAC, because members of Congress have legal authority, and a decision made within a HAC subcommittee might result in the expenditure of a substantial sum of money. Subcommittee votes in the HAC had greater direct consequences than a decision made with an NCPP committee. On the other hand, members of Congress serving on the HAC faced the prospect of interacting together for a number of years, and some members had already served with the same other members on a subcommittee for a number of years. It is normally true that possibilities for agreement among members of a group increase if members expect to interact with one another on a long-term basis. This factor would tend to promote more agreement among HAC members than among NCPP members, who expected the project to terminate in a year or so.

Both the HAC and the NCPP were social systems that behaved as if designed to produce agreement on controversial political issues. Both were very successful. They pose lessons for consensus-building processes in other institutions, such as legislative committees or contending groups within a city or an issue network.

CONCLUSION

The similarities of structure between the HAC and the NCPP point to one way to organize consensus building among cooperative pluralists. In particular, leaders facilitate consensus by propounding appropriate norms of interaction. As in Fenno's HAC and Muir's legislature, the norms of the NCPP were designed to enable participants to educate one another about technical issues and the various interests tied to these issues. The twenty-three leaders in the plenum, however, learned more than the other NCPP participants, because only the leaders discussed all the NCPP issues in a formal process. But since everyone discussed a range of issues in at least one public policy area, in Mill's sense the NCPP was a successful citizenship process among interest group leaders.

The NCPP experiment limited the types of participants. Yet even after restricting the participants to business executives and environmentalists many intense conflicts over issues arose. Wessel's Rule of Reason provided a system of communications in which participants educated one another about interests held in common. Leaders of the NCPP initially were pluralists in their outlook, believing that a coalition of groups could control government policy. Everyone in the NCPP believed that environmentalists provided significant countervailing power to business, and NCPP leaders thought that a coalition of these two powers could control governmental policy. The participants were successful in the first phase of the NCPP experiment—reaching numerous agreements on nontrivial issues.

6
The Platform

Given that the NCPP had evolved a social system well adapted to producing agreement among normally opposed representatives of business and environmentalism, what more precisely was the content of *Where We Agree?*[1]

According to all my interviews and conversations within the Washington policy-making community, everyone, participant and observer alike, considered the content of the NCPP's platform to be significant. (The only exception was the deputy secretary of energy who had not read the platform.) The NCPP actually exceeded the expectations of its leaders in regard to the scope of the eventual agreements. A Resources for the Future energy study project, a Harvard Business School energy study, the *New York Times*, *Fortune*, *Business Week*, the League of Women Voters, the Office of Technology Assessment of the U.S. Congress, and President Carter's Commission for a National Agenda for the Eighties were all impressed by the scope of the NCPP's agreements.[2] Critics almost always attacked the NCPP on other grounds—specifically that the government did not implement its agreements.

The major criticism of the NCPP's platform is that it constituted "cooptation" of one set of interests by the other. For instance, Louise Dunlap of the Environmental Protection Center, Richard Ayres of the Natural Resources Defense Council, and political scientist Douglas Amy viewed the NCPP as an instance of one-sided environmentalist concession to business interests.[3] On the other hand, strong advocates of deregulation, critical of the environmentalist movement, were sure to be displeased by the NCPP's support of the general outlines of the strip-mining regulation act and for public funding of research for environmentalist testimony at permit hearings. To a great extent, one's judgment of "co-optation" depends on how one views the need for governmental regulation of business amidst possibilities of environmental degradation.

Another criticism was expressed by Dunlap, when she observed in an interview that a process similar to the NCPP might be useful if conducted by different people in a different situation. "Maybe this type of mediation would be a good thing in the right situation. It should involve the people who can deliver the political backing for the compromises. It should be on a type of issue on which it can work. A local issue, for example. Or it might be an issue on which the lines of battle haven't been drawn. Some such issues have been discussed in Congress. For example, coal R&D."[4] Richard Ayres's own organization reached a similar conclusion, and thus the NRDC later took part in regulatory negotiations on the regulation of PCB pollution (see Chapter 8). Either environmentalists or free-market advocates, then, might conclude that the substance of *Where We Agree* was generally mistaken, but support the use of regulatory negotiation or similar mediation in another situation regarding different policy issues.

WHERE WE AGREE: SUMMARY

The agreements can be categorized by the five major committees plus another ad hoc committee: mining, air pollution, transportation, energy pricing, fuel utilization and conservation, and emissions charges. In short, the mining committee presented a detailed platform to regulate mining, characterized by a trade-off of environmentalist recognition of the need for flexibility in administering the Surface Mining Control and Reclamation Act of 1977, in return for the industrialists' explicit and detailed statement that many aspects of coal mining require regulation by government.

The representatives of the coal industry agreed that coal mining imposes costs on the environment and on the general public, and that government regulation is necessary to ensure that the mining industry itself pays these costs through more expensive procedures (e.g., building sediment ponds to prevent the sedimentation of streams).

A wide variety of harmful effects are discussed in *Where We Agree*: the devastation of the surface through strip-mining, acid drainage from the mines, sedimentation of streams, the disruption of underground water tables, the leaching of minerals from the earth and their deposit in harmful quantities downstream, the subsidence of the earth above abandoned mines leading to expensive property losses, uncontrolled fires in unworked mines, the competition for scarce water resources in the West, the pollution of the air by the production of dust, the nerve-wracking effects of surface and underground blasting, damage to wildlife and plant species, the destruction of archaeological sites, the creation of the social pathologies of boomtowns, and so forth. *Where We Agree* reminds us why medieval artists often depicted Hell as a mining operation. And even with all this, the NCPP by intention did not treat the issues of the health and safety of mine workers.

In return for the industrialists' acknowledgment of the many sins of mining, the environmentalists conceded a need for flexible administration of governmental regulations of controversial strip-mining practices. They acknowledged a certain relativity to the major arguments against strip-mining. Under certain circumstances, the environmentalists would permit mountaintop removal mining in the Appalachian area, if it were done on an experimental basis, with controls for acid water runoff. Such an exception might be permitted if cheap land were needed to build a subdivision for workers' housing, for example. Anti-strip-mining activists, on the other hand, were angered by this interpretation of strip-mining regulation.

The mining committee of the NCPP also proposed a trade-off between administrative flexibility and public funding for public interest groups:

> Each of the following two recommendations is conditional upon the other:
>
> 1. Regulations (under SMCRA) should be full, unequivocal and definite, but sufficiently flexible to allow for different regional requirements.
> 2. To insure that the environment is being adequately protected, citizen groups should be provided with a reasonable amount of funding to monitor and evaluate the regulatory process and mining operations.[5]

Perhaps the most important proposal of the air pollution committee was another trade-off: in the siting of coal-fired electric power plants, environmentalists would agree to a one-stop permitting procedure if they could get public financing for their testimony and research. Clearly the backing of electric utilities and other businesses would greatly facilitate such public financing. The air pollution committee also proposed that under limited conditions tall stacks be allowed as a method of emissions control on certain old power plants, rather than the alternative of forcing the installation of expensive limewater spray and other "scrubbing" devices to remove sulfur dioxide. The committee stipulated that such older plants with tall stacks must still meet the "ambient air quality standards" for sulfur dioxide and that power companies should use the capital saved by the retention of tall stacks to install scrubbers in new coal-fired plants. Another proposal from the air pollution committee was that coal-fired plants should be built in the vicinity of the users of the electric power rather than at the "minemouth," with the electricity transported by long-distance power lines.

The report of the transportation committee would gladden the heart of the academic economist. Generally speaking, half the price paid for coal in the United States is transportation costs, thus the close link between transportation policies and coal policy. The transportation committee principally argued for governmental deregulation in the transportation area. In particular, the NCPP supported deregulating trucking (trucks are widely used for short hauls of coal), eliminating legal barriers to the construction of coal slurry pipelines, allowing railroads to negotiate long-term contracts for hauling coal, and charging coal barges "user fees" for the navigation of federally constructed river channels. (The lack of such user charges, instituted in partial form in 1978, was widely considered a subsidy to the river barge industry.)

The principal recommendation of the energy pricing committee was marginal cost pricing, another concept favored by many economists. The general practice in the pricing of electricity is to charge the same price for all the electricity used by a customer, even though the cost of an increase in electricity consumed may be much greater because new power plants are more expensive than older ones. In the NCPP's marginal cost pricing plan, the customer would be charged the higher costs of the additional units of electricity in the case of an increase of power usage, but the overall amount charged by the utility to all users should be kept constant.

Further, the proportions charged to various types of users of electricity would remain constant, thus distinguishing the NCPP's position from those calling for an increase in the lower rates often charged to industrial users of large amounts of power. Marginal cost pricing would charge the consumer more per unit, the more power one used, thereby serving as an incentive to conserve energy. The committee opposed charging special low rates to poor consumers on the grounds that rates should reflect the actual costs of delivering electricity, though many committee members supported a program of grants to the poor to pay for increased energy costs.

The fuel utilization and conservation committee offered the fewest recommendations. It supported a flexible administration of mandated coal conversion of oil/gas fired plants to coal-fired electrical production. It advocated a more aggressive public policy to further "cogeneration," constructing electric generation plants at the site of industrial steam heating and similar facilities. (The same steam can heat a building and drive a turbine to make electricity.)

A special committee proposed charging an emissions tax for air pollution from "stationary sources" (large industrial plants). Another idea favored by most economists but not yet introduced into public policy, an emissions tax would be a more flexible instrument of regulation than a binding rule would be, because the polluter would decide what degree of pollution might be efficient in terms of his costs for pollution control, other production costs, and the pollution tax imposed by government. Theoretically, some additional pollution would be emitted if pollution control were expensive in terms of other economic goods, while less pollution would be emitted if it were particularly cheap to control it.[6]

The proposals of the NCPP can be summarized in three points. (1) *Where We Agree* sets forth a detailed discussion of how coal mining—both surface and underground—might be regulated. (2) *Where We Agree* argues that simplified licensing procedures, requiring less time and combined with public financing of "public-interest" intervenors, is in the general interest. (3) The NCPP's proposals contain numerous suggestions that public policy should reflect economic theory. The adoption of marginal cost pricing for electricity and the idea of emissions taxes are two notable examples. Deregulation in the transportation sector is another important example; few economists support subsidies for the river barge industry or legal prohibitions to the organization of coal slurry pipelines.

The proposal incorporating simplified licensing procedures and public financing for "public interest" intervenors in licensing hearings was not seen in 1978–1979 as a preeminent proposal of the NCPP. But it was one of the three parts of *Where We Agree* to be specifically turned into a proposed bill and introduced into the legislative process.[7] This licensing trade-off proposal is particularly important, as we will see in the next section.

The NCPP's platform on cogeneration as well as its emphasis on the importance of considering the hydrology of an entire river basin, rather than limiting oneself to man-made political boundaries, in designating strip-mining regulations have assumed additional importance in retrospect, because these two items were incorporated into public policy (see Chapter 7).

The NCPP itself described its findings differently in ten principles stated in the summary of *Where We Agree*.[8] One "principle" is a residual category ("certain important gaps in present policy should be filled"),[9] five are simple statements of economic principles, and the remaining four concern the substance of coal-related issues.

The five economic principles are as follows: One, policies that require the beneficiary to pay the full cost of using "public goods" should be encouraged. Two, the government's role in the economic regulation of energy markets should be reduced by placing greater reliance on free-market mechanisms. Three, dollars spent on pollution abatement should be directed to achieve the maximum reduction in pollution. Four, public policy should discourage the uneconomical use of energy and prices should reflect the marginal cost of the additional units of energy consumed. Five, energy policy and social policy should be addressed separately. The last point means that energy use by the poor or by other groups should not be directly subsidized, although "income transfers might . . . be considered. On the other hand, policies which artificially depress energy prices cause distortions of market and investment decisions, lead to higher rates of inflation, and eventually produce an even greater adverse impact on those of low income."[10]

Of the remaining four principles, one is common to most such commissions: "Policies should encourage the implementation of new technology."[11] Another asserts regulatory flexibility in strip-mining: "Mining should be encouraged in areas where ways to avoid significant environmental problems have been demonstrated and discouraged in areas where this has not been done."[12] A third describes the trade-off in compromises in utility siting: "Means should be established to expedite decisions regarding utility plant siting, with safeguards to ensure that appropriate alternatives are considered and that all concerned parties are effectively represented."[13] A final, somewhat surprising, basic principle is discussed below: "Major coal-burning power plants should be located, to the extent practicable, in the general area where the energy is to be distributed."[14]

MUTUAL RECOGNITION OF LEGITIMACY

The cooperative social processes described in Chapter 5 encouraged participants in the NCPP to recognize a considerable legitimacy in the positions of the two sides. This recognition, in turn, was a basic step toward the formulation of *Where We Agree*, enabling both sides to back off from their ideally preferred positions and to compromise somewhat with the views of the other side. Within the special negotiating conference of the NCPP, it was easier to compromise ideal positions, because lobbyists from both sides could continue to advocate their preferred positions outside the NCPP in Congress and elsewhere.

Business executives, after all, are citizens and breathers of the common air, and thus they recognized that someone must represent the interest of the environment in the American interest-based political system. Profit-making business could not fulfill that role, nor did business participants relish the idea of government being the sole advocate of environmental interests. Accordingly, business participants came to recognize the legitimacy of environmentalist lobbies, especially as, in the NCPP, they saw environmentalists behave rationally and civilly, respecting the norms of economics, science, and voluntarism that were shared by both sides. It was but a short step from recognizing the legitimacy of environmentalist lobbies to accepting partial public funding for such lobbies, in return for a quid pro quo.

Environmentalists recognized that someone must run the economy and that some mining and electricity production is part of the modern world. And while theoretically they saw no need to increase these activities, the majority of NCPP's environmentalist participants had no objection to the private enterprise economy—after all, it might prove difficult to stop increases in coal mining or electricity production if these were controlled by government bureaucracy, as in European countries. Environmentalist participants recognized that privately owned business, in conjunction with market-set prices, would avoid subsidizing energy prices, thereby disciplining public consumption of energy.

Environmentalist participants were not so much opposed to private business and profits, but to the wasteful consumption of energy. Accordingly, environmentalist lobbyists were willing to accept the legitimacy of the role of coal-company or private-utility executives. And again, the acceptance was made easier because the business side shared the norms of economics, science, and voluntarism. It was then a short step for the environmentalists to accept the necessity for utilities occasionally to construct new power plants and for coal companies occasionally to develop new mining operations.

In effect, both business and environmentalist participants advocated a two-branched political strategy for their own group. Business participants generally recognized the need for some air-pollution and strip-mining regulations, but ideally would have preferred major amendments in the law to ease regulations. At the same time, accepting the likely continuation of the laws without major amendment, they accepted the need to negotiate with environmentalists to gain flexible administration. Coal industry lobbyist Harrison Loesch worked to modify the 1977 SMCRA by delegating much of the regulation writing to branches of state government. At the same time, knowing the amendments were unlikely to be passed, Loesch negotiated with environmentalists in the NCPP. Similarly, some of the environmentalist participants in the NCPP ideally would have opposed any strip-mining of coal in the Northern Plains area, but realizing that such advocacy would fail, settled for negotiating with coal company executives in the NCPP to encourage the mining of coal in ways least harmful to the environment. Yet while participating in the NCPP such groups as the Sierra Club were still free to persuade the general public to follow policies that would greatly cut back energy consumption.

TRADE-OFFS TO EXPEDITE DECISIONS

The idea of trading off funding for environmentalists for an agreement to expedite hearings actually is a restatement of the core of the Rule of Reason, which enjoins a full presentation of information by all sides and criticizes the use of strategies to conceal relevant information on some issues. A direct implication is that each side of an issue should have enough resources to do the research needed to develop the factual support for its case. Another direct implication of the Rule of Reason is that political opponents should not manipulate institutional rules, such as intentionally stalling decision-making procedures, to prevent public discussion of an issue.

Presently, because environmentalists oppose increases in the use of energy and any major degradations of the environment, and because environmentalist lobbies and litigation groups typically do not have much money, they follow a strategy of delay and obstruction to prevent the development of new plants and mines. As the reader may know, the construction of new mines or power plants may require twenty different permits, from federal, state, or local authorities. Separate laws regarding strip-mining, air pollution, environmental impact statements, nondegradation of watersheds, and so forth each require a permit, which is granted only after an application and hearing process before the relevant board or government agency. Given their lack of money in relation to corporate opponents, environmentalists pursue a prime strategy similar to that of guerilla war—fight a skirmish in one permit proceeding, fight another in a second and a third, expect to lose the direct legal skirmishes because of inferior legal resources, but fight a strategy of skirmish, appeal and delay, in one arena after another, until the costs of the project have increased (particularly in times of inflation) to the degree that it is no longer worthwhile for business to continue. One now normally expects environmentalists and citizens' groups to use such a strategy to block the construction of new power plants and mines, as well as other large projects. But environmentalist lobbies should not be criticized for this strategy. It is only a rational use of their limited financial resources. They are only following the incentives established by the wider political system.

This is the situation that NCPP's participants sought to correct. Why not give the environmentalists enough money to conduct research and to build their legal cases, if they would agree to give up their skirmish-and-delay guerilla war strategy within the legal system? Specifically, environmentalists would accept a consolidation of the various permitting processes insofar as possible into one major process. Within such a consolidated process, the environmentalists could better present the facts of their case, since they would have received public grants to prepare these facts and present them effectively. Hearing procedures might then more closely resemble Wessel's Rule of Reason, rather than "sportsmanship" or guerilla war. Governmental commissions and agency chiefs would then be able to make better decisions, since they would receive more balanced information. Society would profit because decisions could be made more quickly. Of course, the consolidated permitting hearing might make a relatively speedy decision *not* to build, thereby rapidly eliminating a faulty policy. Business could then move on to propose a better-designed project.

As part of this trade-off, state governments would be encouraged to devise plans for long-run land use and power production. State government might then require private business to notify it of new projects as early as possible under the strictures of the state government's plan. This notification would give environmentalists and citizens' groups time to mobilize for their participation in the consolidated hearing, rather than have the project and the hearing suddenly announced as part of a business strategy to overwhelm opponents with a quick strike.

On the other hand, joint federal-state lawmaking mandating the consolidated permitting process might somewhat restrict standing-to-sue by groups objecting to the outcome of the hearing. The procedure would be to notify, as early as possible, all groups and business that might be affected by the proposed project, and, having given them plenty of notice and the possibility of public funds to prepare testimony, then restrict the right to sue to block the outcome of the consolidated hearing.

The leadership of the NCPP added another item, user proximate siting, to get environmentalists to accept this trade-off. This was the enunciation of the general principle that electric power plants should be built in the area of those using the power, rather than in the area of the mining process.[15] NCPP-sponsored legislation directed that new power plants should be built in the state whose residents used the majority of the power (i.e., no more transmitting power to California from plants in Arizona, Utah, or New Mexico).[16] This principle was seen as incidental to the norm of economics—it is preferable that those who benefit from some production process also pay the costs of such a process. (If a city benefits from the electricity from a coal-burning plant, then that city should pay for the pollution costs of that plant.) In addition, environmentalist participants recognized that if the principle of user proximate siting were to be adopted in practice, it would serve as an incentive for cities to reduce their consumption of electricity, as the residents suffered the costs of pollution induced by increased production.

The consolidation-hearing and public-funding trade-off recommendations were made by the air pollution task force.[17] The strip-mining committee also recommended a similar, but somewhat more limited trade-off. It advocated a trade-off of business acceptance of public funding for groups lacking resources for preparation of siting hearings with environmentalist acceptance of the principle of flexible administration of the strip-mining law. Given adequate funding, environmentalist groups might present their case at hearings concerning the granting of exceptional permits under the law. The strip-mining group of the NCPP backed into a sort of user proximate siting position by recommending that: "Mine-mouth coal conversion facilities should not be encouraged in the Northern Plains and Rocky Mountain coal regions and prime agricultural areas of the Midwest."[18] This statement seemed to apply to coal-burning power plants as well as to coal conversion plants (e.g., coal gasification). This position implies that power plants would be built closer to the users in California, Denver, Texas, and the urban Midwest. The strip-mining group stated that this recommendation "is consistent with the facility siting recommendation of the Air Pollution Task Force."[19] The strip-mining task force did not, however, clearly recommend consolidated permitting hearings or long-term plans by state governments for the development of strip mines.

The consolidated-hearings and public-funding trade-off was converted into legislative language by the NCPP's staff in 1979. Subsequently, its provisions were introduced as a bill by Democratic Representative Don Pease of Oberlin, Ohio, as H.R. 1430 in the Ninety-seventh Congress on January 28, 1981. Congressman Pease, however, restricted the language of the bill to coal-fired power plants, rather than energy facilities in general, as the NCPP had advocated. But coal-fired power plants had been the NCPP's principal concern.

Pease's first bill and other bills introduced into subsequent Congresses were not even voted on in committee. The NCPP-advocated trade-off still must be considered a concept out of the legislative mainstream, largely owing to the increase in confrontation politics in environmental policy initiated by Reagan, Watt, and Gorsuch. In a time when the political leaders of business supported major cutbacks in funding for the EPA and the strip-mining enforcement agency, the Office of Surface Mining in

James Watt's Interior Department, environmentalists could not be expected to show much interest in legislative proposals to make deals with business in the public interest. Furthermore, since the mid-1970s a leading business lobby, the U.S. Chamber of Commerce, has opposed the concept of public funding for environmentalist and citizens' groups, posing an important barrier to the passage of a bill like Pease's H.R. 1430. But there has been a deescalation of such confrontation between business and environmentalists during the Bush Administration. Consequently, passage of a measure like the NCPP's trade-off might be politically feasible in the 1990s. In Washington politics, it is common for innovative measures to be passed ten years or more after their first legislative introduction.[20]

THE NORM OF ECONOMICS

In addition to the Rule of Reason norms, the belief in the applicability of economic concepts shaped the content of *Where We Agree*. Five of the nine basic principles stated in the summary of *Where We Agree* are economic principles. The reports of three of the six major task forces primarily state the value of applying market mechanisms and economic incentives—such reports were issued by the Transportation, Energy Pricing, and Ad Hoc Emissions Charge task forces.

Much of the spirit of the NCPP was evident at a meeting of the plenum attended in September 1979. The meeting was called to order by Father Francis X. Quinn, a professor and labor mediator, who solemnly reminded the assembled industrialists and environmentalists of the content of the group's scriptures: *The Rule of Reason*. Next, Larry Moss, the leader of the environmentalist caucus within the NCPP, addressed the group, reminding them of their mutual dedication to the expansion of market principles within the coal sector of the economy. Moss reminded the plenum that government intervention is justified in order to get the market to reflect the true social costs of coal production, but that he—and everyone else within the NCPP—is opposed to excessive governmental intervention in the coal economy. The symbolism was clear: the message presented by the leaders of the NCPP is that rational, disinterested, public-spirited discussion can produce agreement between disputants by the application of economic principles and concepts.

We are not surprised that representatives of business value the "free market," which is minimally constrained by government. Nor is it surprising that industrial leaders find free-market ideas persuasive as a justification for supporting proposals such as those advocated by the NCPP. But many people were surprised to find the environmentalists in the NCPP using free-market terminology.

The New Deal liberals of the 1930s through the 1960s usually scorned such terminology, which they associated with their opponents' arguments against the Democratic party platform of increased federal action in the economy. However, during the 1970s, partly because of the discrediting of the federal government by the war in Vietnam and Watergate, partly because of the well-known failures of the Great Society programs, many politicians and intellectuals of the center-left became skeptical of governmental action and supportive of free-market values. Environmentalists, however, had not been a very noticeable part of this trend within the center-left. Indeed, environmentalists usually impressed others as using the vocabulary of natural law, moralism, and populist denunciation of large economic and governmental institutions.

However, a trend within academic political economy caught the attention of some environmentalists. During the 1960s and 1970s, economists paid increased attention to questions of "market failure," and of "public goods." These concepts overlapped many of the situations of greatest concern to the environmentalists. "Market failure" refers to situations in which an economic actor imposes "external costs" upon his environment, but there is no feasible way to charge the actor for the costs he is imposing on others without some authoritative intervention. Many instances of pollution have this quality. Similarly, a "public good" is one that is jointly supplied to a group and which cannot be readily appropriated for sale: clean air and water are classic examples. Environmentalists, of course, are often the defenders of public goods, which may be undervalued in the political-economic marketplace.

Economists dealt with questions of market failure and of public goods by discussing how the "true social costs" might somehow be measured and how the "externalities might be internalized" in the cost-benefit calculations of economic actors. Accordingly, some of the more pragmatic environmentalists (such as Larry Moss) realized that the concepts of economics could be used *for* them, in arguing that polluters impose costs upon society, as well as against them. Some environmentalists also were persuaded that economists are correct in arguing that financial incentives to polluters are more effective than governmental fiat in attaining clean air and water.

So the stage was set for mutual agreement between at least some industrialists and environmentalists. Businesspeople were naturally pleased that environmentalist participants in the NCPP talked the language of governmental deregulation, pricing by markets, and disinterest in the divestiture of coal holdings by oil companies. In return, businessmen could not resist the language of "internalizing the externalities," as it applied to coal mining. Agreement on governmental deregulation and a lowering of subsidies in the transportation sector follows readily from the trends just discussed. After all, even in Congress, probusiness conservatives joined Senator Kennedy in pressing for deregulation of airlines, railroads, and trucking.[21] Similarly, both sides in the NCPP could agree on the systems of financial incentives embodied in the emissions tax and marginal-cost pricing ideas.

The recognition of the legitimacy of one another's roles and the development of mutual trust produced a flexible outlook on both sides, which in turn enhanced the tendency of environmentalists to agree to the language of markets and incentives preferred by business. The artificial forum of the NCPP brought out the environmentalists' capacity to see market mechanisms and incentives for what they are—a means to an end that might often be the control of pollution, a "public use of private interest."[22] This recognition might be counted as one of the chief accomplishments of the NCPP mechanism.

Outside of the NCPP, environmental lobbyists have not tended to advocate deregulation and the substitution of incentives for agency commands in the implementation of policy. Yet Harvard professor Stephen Kelman showed in a study of Washington politicians and government executives that a significant number of environmental lobbyists interviewed favored experimentation with the emissions tax approach to pollution control.[23] And surveys by Robert Cameron Mitchell clearly indicate that contributors to environmentalist organizations represent a much broader range of political viewpoints than traditional liberals or radicals, who prefer to expand the scope of the federal government and may be suspicious of market mechanisms.[24] The nation's political forum apparently suppressed the advocacy by environmentalists of deregulation and the substitution of economic incentives for legal orders. Why did this happen?

One reason is that environmentalists who would be willing to experiment with emissions taxes and other such means for pollution control hesitate to support such measures because other environmentalists would be adamantly opposed, leading to disagreements within a lobbying coalition that participants prefer to avoid. Similarly, environmentalists would seek to avoid offending the old-fashioned New Deal Democrats whom they depend on for support in legislatures and as lobbying partners. Such consideration would hold in the case of the issue of deregulation of natural gas prices, to cite an important example. In that particular case, environmentalists attracted to deregulatory concepts were deterred because while a price hike conserves gas, polluting coal or high-sulfur oil might be substituted for the exceptionally clean-burning natural gas.

In this respect, the Washington polemics in the 1970s over energy and environmental policy were misleading and probably hindered the development of effective policy. When the environmentalist movement reached the legislative arena around 1969, there was a general rush to support the goals of so many upper-middle-class voters. Business lobbyists were surprised and disoriented until around 1974,[25] when they realized that the most effective arguments to slow or block environmental legislation were the combination: (1) environmental legislation increases the span of federal regulation; (2) such legislation is not informed by the realities of economics and is needlessly costly; (3) such legislation increases the costs of production and is inflationary, contributing to a severe national problem. After deftly accusing environmentalists of lack of concern for economics, leaders of the business community had difficulty recognizing that many environmentalists were, in fact, privately supportive of market mechanisms that would reduce the consumption of energy and thereby reduce pollution. Environmentalists, in turn, chiefly opposed the business onslaught by continuing to demonstrate the bad effects of environmental degradation, rather than turning aside the business attack by advocacy of deregulation, emissions taxes, and marginal cost pricing of electricity. In this sense, the NCPP was not realistic. But it was more creative and more conscious of the general interest than the real policy forum.

THE NORMS OF SCIENCE

When the Mining Task Force first met in February 1977, the Surface Mining Control and Reclamation Act (SMCRA) had yet to pass, although it was to be signed by President Carter a few months later. However, the general outline of the proposed act was known to NCPP members, and business participants were in general opposed to the bill, while environmentalists favored it. As the Mining Task Force continued to meet in 1977, however, the consensus-building processes had an impact. Each side recognized a basic legitimacy in the other's positions, and both backed away from their ideal preferences. The business side was ready to acknowledge the legitimacy of some federal regulation of strip-mining, if such regulation were flexible and recognized certain arguments for increased production. Environmentalists, on the other hand, valued such support from coal-company executives and other business representatives for strip-mining regulations. In the NCPP, both sides agreed that a joint business-environmentalist platform would be a common gain, if this platform were based on business's acknowledgment of the legitimacy of rules similar to those passed under SMCRA, and environmentalists acknowledged the need for some flexibility in rule writing in the interests of inexpensive production.

The developing agreement was enhanced by the norms of science, accepted by both sides of the mining committee. Such norms obviously overlap with the Rule of Reason beliefs, in which participants are enjoined to make all relevant information public for the sake of objective discussion. But mining committee participants went even further. They did what amounts to collaborative research on the question of what flexible interpretation of strip-mining regulations would mean. Accordingly, they hired two full-time staff experts and, as a group, visited surface mines in Pennsylvania, West Virginia, Ohio, Illinois, Texas, and Montana. The committee spent about eight days in site visits.

The mining committee thus followed some of the basic norms of science. They gathered data about mining. They left their conference room to get a firsthand look at the "evidence." They in effect attended lectures, given by mine executives or environmentalists at the mining sites. The committee resolved to be inductive; a first premise was that conditions might differ in various environments. They thus analyzed the problem of strip-mining regulation into parts (regions) and examined the evidence about each region separately.

Facts cannot compel agreement among parties with different values. But the strip-mining committee demonstrated that it is another matter if such "facts" are jointly discovered by opposed groups on a joint research mission. If there is a spirit of goodwill among participants, and if there is a process of consensus within the group, such "facts," new to the participants, become a new, jointly held interpretation of reality, a common viewpoint among normally opposed parties. In this sense, the NCPP's norm of science led to a search for new "facts," which became the basis for an assertion of common interests.

The NCPP's investigation of regional variation in mining was based on geological variation and the prediction that there would be no great increase of surface-mined coal from the Northern Plains region. During the early Carter administration, many experts predicted that the United States would need to greatly increase its coal production, perhaps doubling it within a period of twelve years or so. In fact, this need did not arise, largely due to a big decline in the rate of increase of electricity use and to the lack of need for "synfuels," such as gasified coal. But the prevailing wisdom of 1977 was that strip-mining in the Wyoming/Montana region would greatly increase, and the NCPP believed this would not occur. This belief enabled business to agree with environmentalists on the value of restricting Northern Plains production for environmental reasons, since participants in the NCPP believed that most of the projected increased production from this region would not in fact be produced anyway. The business and environmentalist sides did disagree on the degree of restriction of strip-mined production from that area, however.

In addition, both sides in the NCPP agreed on a regulatory priority to control acidic water runoff from Appalachian mines, a problem that was not particularly important in the West. In the arid West, on the other hand, wasteful water usage in mining needs to be closely regulated. In the northern portions of the midwestern coal fields of Illinois/Indiana, the main regulatory priority should be to prevent the destruction of rich farmland and to force coal miners to restore the earth to its former fertility, or not to mine at all. Regulatory inflexibility, about which business complains with considerable justice, occurs when Northern Plains strip-miners must follow detailed regulations concerning acid runoff which is not much of a problem, while Appalachian miners are carefully limited in the amount of water to be used for mining operations.

The NCPP set forth a de facto statement of regional priorities in coal mining. In general, the NCPP preferred the extension of existing mining in the deep mines of Appalachia and in the Southern Illinois–Indiana basin. The two sides agreed in stating that the most arid regions of the West should not be strip-mined for coal, because of the impossibility of reclaiming the land due to difficulties in replanting. Both sides agreed that a low priority be given to extending the strip mines in the northern Wyoming, southern Montana region, although they disagreed over the environmentalists' proposition that a form of zoning be created, which would restrict coal mining for the time being to the regions around Gillette, Wyoming, and Decker, Montana.

The business participants of the NCPP subscribed to very extensive federal regulation of mining, something they had opposed before the NCPP. The Mining Task Force report occupies the entire second volume of the unabridged *Where We Agree* and consists of 477 pages, including appendices. It is a detailed statement of the problems of coal mining, especially of strip-mining, and makes about 150 recommendations for public policy. More than 100 of these recommendations imply types of regulations that might be drawn up under the strip-mining and other environmental control laws. At the time of writing, the SMCRA of 1977 had just passed and regulations had not been drawn up. Other recommendations offered suggestions on agenda-setting or advocated conducting certain types of environmental impact research. In other words, many of the recommendations called attention to an environmental problem of coal mining and essentially stated that more attention should be paid to the problem, without explicitly advocating governmental regulations.

But the 100 or so regulatory recommendations implied a lot of regulation—no doubt more than a thousand pages of rules actually drawn up. (Of course, under the SMCRA many of the specific rules would be drawn up by state governments.)

The NCPP advocated regulation of almost every negative environmental impact of coal mining, including strip-mining, that might concern the environmentalist. Areas of regulation included restoration of the contours of the land and replanting the soil with vegetation in the case of strip-mining, control of acid runoff from deep mines (in which water washing through pyritic rocks produces an acidic liquid similar to acid rain), general control of the effects of mining on the water table and on the hydrology of river basins, prevention of landslides from piles of earth removed in mining, control for soil subsidences into mines (a major problem in deep-mining areas), control of the creation of dust in arid regions of the West, concern for the stubborn problem of unchecked fires in old mines, regulation of mine blasting which unsettles nearby inhabitants, designation of responsibility for problems in abandoned mines without owners, establishing plans to drain mines filled with acidic water, the protection of cultural and historical artifacts such as Indian relics, "critical winter range for antelope should be maintained and restored,"[26] and more.

CONCLUSIONS

The norms of the NCPP, the cultural beliefs that promoted solidarity and consensus, also shaped the content of its platform. The Rule of Reason norms not only shaped the mode of negotiation, but also prefigured the trade-off in which both sides gave up institutional strategic advantages for the sake of timely decision making, reliant upon better information. The concepts indicated by the norm of economics provided a way for both sides to think about possible unrecognized common interests. The norm of science prescribed analysis of a problem into its parts and then a joint search for new empirical evidence bearing on the problem. The new joint interpretation of the "facts" could then be subjected to Rule of Reason discussions and viewed in the light of economic concepts.

These results point to a generalization applying to negotiating bodies: the norms of such institutions may shape their policy output. Rules of negotiation may be embodied in such policies, as in the case of the Rule of Reason and the trade-off to expedite decisions. Similarly, jointly held conceptual frameworks about policy (e.g., economic concepts) or ways to gather new information may be embodied in the policy output of a negotiating body.

My analysis of the NCPP resembles the classical studies of the House Appropriations Committee in this respect, as it does in the modes of reaching consensus. Relying upon Fenno's pathbreaking observations on the HAC, Aaron Wildavsky demonstrated in *The Politics of the Budgetary Process* that the participants in the HAC followed the norms of incrementalism, and that these norms shaped the budgetary output of the committee. For instance, in hearings subcommittee members of HAC usually focused their examination on proposed changes in the budget from the preceding year to the present year. The members of Congress almost never considered the entire budget of an agency from "the zero base." This norm of incrementalism was regularly reflected in the HAC appropriations figures for the budgets of most federal agencies. Such appropriations typically varied incrementally from one year to the next—the groups negotiating norms shaped the policy output of the group.[27]

However, group norms may shape the platform and the activity of the group in a way unhelpful to attaining its goals. In the NCPP's case, the norm of voluntarism, while contributing to the group's own sense of unity, impeded implementing *Where We Agree* in public policy. The norm of voluntarism kept governmental negotiators out of the NCPP's processes. But, as I discuss in the next chapter, the NCPP's leaders failed in their subsequent attempt to get federal regulators to follow the NCPP's recommendations, a failure that was made more likely by barring governmental personnel from the NCPP in the first place.

In general, norms of cooperative pluralist negotiating bodies may sometimes have an "us" versus "them" character—"we" serve the public interest better than "they." But negotiators should consider whether they would subsequently need the cooperation of "them." If so, establishing such a norm is likely to undermine the implementation of the negotiating institution's platform.

The NCPP's norms of the Rule of Reason, economics, and science might sometimes be appropriate for other cooperative pluralist negotiating bodies. One application might be to regulatory negotiation (see Chapter 8), in which government officials and opposed interest groups attempt joint development of consensual regulations on some rather specific topic.

For the individual participants in the NCPP, the three norms provided a basis for learning about policy issues. The two volumes of *Where We Agree* show that the NCPP was a learning process for the participants. When the project began, Decker and Moss had an idea that the conferees might eventually agree on some combination of economic and environmental principles. But neither they nor anyone else at the beginning of the NCPP had a remote idea of the subtleties and complexities of the eventual *Where We Agree* statement. The various combinations of economic and environmental principles were extraordinary, and even today seem quite unusual. The general and specific points of the joint statement were developed in the committees, so most of the active participants in the subcommittees gained a new perspective on policies in areas of their own vocational and political interest. *Where We Agree* is probably less striking to academicians, because it reads like a document put together by a group of, say, business administration professors meeting with a group of environmental scientists. But that is the point—the processes of the NCPP converted a group of businessmen, lawyers, and lobbyists into something like a group of researchers, working on common problems.

Several leaders of the NCPP told interviewers in 1982 that they had become accustomed to contacting participants on the other side of the NCPP to broaden their viewpoints about new environmental or energy issues.[28] Of course I cannot prove that participation in the NCPP has continuously affected the political behavior of participants for a decade or more. The aim of this study is just to show the possibility for political adversaries to learn from one another and how institutions can be developed to further this end. If adversaries come to an awareness of common, as well as conflicting, interests, they achieve a valued goal in Mill's philosophy of citizenship and in the normative perspective of cooperative pluralism.

7
The Failure of Lobbying

The history of the NCPP reveals two major surprises. The first surprise has just been presented: adversaries were able to agree upon such a broad platform. But the second surprise was not so pleasant: the businesses and public interest groups represented in the NCPP would not endorse its consensus platform. Furthermore, the NCPP's relationships with the Congress were generally ineffective.

Recent interest group theory finds that policy-making is normally tripartite, with producer groups, countervailing groups, and government agencies each having an independent impact on policy outcomes. In such theory, one must consider the effects of government agencies and not regard coalitions of interest groups as the sole causal agent in determining policy. But in its norm of voluntarism and reliance on the formation of a grand coalition of groups to lobby for its platform, the leaders of the NCPP ignored the autonomous role of government. Since they were unable to form the grand lobbying coalition of environmentalists with the coal and electric utility industries, the NCPP was left with little to work with in attempting to enact its platform.

THE IMPLEMENTATION PHASE

In its own terminology, the "first phase" of the NCPP ended with the publication of *Where We Agree* in February 1978; the "second phase" consisted of trying to implement these agreements plus negotiating unsolved issues such as the leasing of western coal lands. The second phase ended with the last meeting of the plenum of the NCPP in March 1980; after that, the committees of the project did not meet, although some windup activity was conducted by the staff at Georgetown's CSIS and by some of the project's leaders in informal meetings with executive branch officials. The final report of the project was completed by the spring of 1981. In addition, *Why They Agreed: A Critique and Analysis of the National Coal Policy Project* by Francis Murray and J. Charles Curran was written in 1982 and published by the CSIS in November 1982.

The implementation function was conducted by Murray and staff assistants at Georgetown CSIS. A special appeal was made to foundations to finance this "implementation" phase; one reason the word "lobbying" was not used is that since 1969 foundations were forbidden by the Internal Revenue Service from "lobbying," although in a very limited, legal sense of the term (soliciting a legislator's vote on a specific bill). In all, the project implementers at Georgetown spent about $400,000 over three years, most going to salaries and rent, as is typical of lobbying organizations. In addition to the salaried staff at Georgetown, Larry Moss, Jerry Decker, Mac Whiting, and some other project leaders volunteered to contact executive branch officials, members of Congress, and the press on behalf of the NCPP.

Implementation activities were put in Murray's charge with a general understanding that he would coordinate the activities of other participants of the NCPP who wished to lobby on behalf of the project. But few requests were made of participants who did not volunteer to lobby; apparently most of the NCPP's negotiators assumed that implementation "was being taken care of by the office in Washington."[1] A year after the publication of *Where We Agree*, it became apparent that the lobbying effort was not meeting the expectations of most of the participants, some of whom were surprised and quite disappointed. These participants, however, did not do much to lobby their own environmentalist organizations or businesses and trade associations in behalf of the NCPP's platform, because they all shortly realized that such efforts were hopeless and persistent effort on behalf of *Where We Agree* would only annoy bosses and governing boards.

PRESS RELATIONS

Relationships with the press were quite successful during the period of February to April 1978, immediately following the publication of *Where We Agree*. Many leading newspapers and trade periodicals carried stories about the project, and their overall tone was favorable. The NCPP's chief successes were perhaps a laudatory article in *Fortune* magazine (many copies of which were distributed by the project itself),[2] a front-page story and favorable editorial in the *New York Times*,[3] and coverage by the UPI wire service. The favorable tone of coverage is indicated in the first two paragraphs of the *New York Times* story:

> After 13 months of debate and study, a group of leading environmentalists and industrialists announced today that, to their surprise, they had agreed on more than 200 steps that could help the nation shift from oil to coal as a prime source of energy in ways that are both environmentally tolerable and economically sound. Although the group's findings are in no sense binding on either industry or regulatory authorities, the participants in the National Coal Policy Project said that their work had been "devoid of the extremism" of the past. They expressed hope that it could help in implementing present mining and environmental laws and in reconciling wide divergencies of opinion on coal policy.

Certainly, the average reader of the *New York Times*, not connected to either an environmentalist lobby or to the coal industry, would react favorably to this news story, which seems to recount a surprising breakthrough in the national interest of alleviating America's energy shortages.

Major news stories about the NCPP usually did mention somewhere in the last half of the piece that a number of environmentalists were critical of the project. The sources of criticism were always Dunlap and Ayres. Indeed, veteran lobbyist Dunlap made sure her point of view would be covered when she appeared at the press conference announcing *Where We Agree* and distributed her own statement criticizing the NCPP. The *New York Times*, for instance, reported on an inside page:

> Their work immediately came under sharp attack, however, from environmentalists, who had declined to participate. These critics

> charged that some recommendations would weaken provisions of strip mining and pollution control laws that had taken years of legislative conflict to enact. The critics also protested that the experiment in reducing conflict to society had excluded interested parties.[4]

Other press coverage also indicated briefly the views of Dunlap or Ayres. But the incorporation of Dunlop's criticism into early press coverage did not damage the project's image, in my own view. Part of the attraction of the NCPP derived from its status as an apparently rational, disinterested attempt by parties of goodwill to resolve their differences in the national interest. Dunlap's statements reminding the reader of environmentalist-business conflicts actually underline the importance of the NCPP's message. Moreover, the fact that someone important opposed the project indicated that its report dealt with significant issues and was not merely a bland statement of abstractions.

After the publication of and public reaction to *Where We Agree*, the NCPP did not make much news, as it is ordinarily defined by the press. Consequently, the project did not get much press coverage after the spring of 1978, with a few exceptions. Local press might report a meeting, speech, or seminar conducted by the NCPP. The *Denver Post* and the *Rocky Mountain News* (Denver) lauded the project in articles dealing with the need to resolve disputes that retard energy production, a matter of exceptional interest in a city that considered itself to be the nation's energy capital.

FAVORABLE ACADEMIC REACTION

In 1979, the NCPP received considerable attention and favorable treatment in the two most prestigious general assessments of America's energy situation that appeared at the time: the Harvard Energy Study, published as *Energy Future*; and a study by the Washington research institute, Resources for the Future (RFF), published as *Energy and America's Future*.[5] Both studies were being written in February 1978, when *Where We Agree* was published, and thus they reflect the good initial publicity received by the project.

These two studies indicated that among leading scholars of America's energy problems, some were very impressed by the NCPP. Such supporters of the NCPP recognized that the project was not suggesting solutions to most of the conflicts in the coal area, one of the critics' major points. Hence it did not matter to the supporters that the NCPP did not represent all groups in the coal area, i.e., the NCPP was not perceived as an attempt to resolve *all* disputes about the production of coal. Supporters referred to the NCPP as something novel and constructive, a first step, a good example, and so forth. Supporters used the language of hope, not the language of completion. The prestigious supporters of the NCPP never referred to failures of implementation, another charge against the project, perhaps because they urged emulation of the NCPP and references to this major problem would weaken their case.

To turn to the Harvard Energy Study, a page of praise was given to the NCPP, which is discussed prominently at the conclusion of the chapter about coal:

> About the same time [as the coal strike of 1977–1978], however, a more promising way to cope with another of the constraints [on the abundant usage of coal]—the environmental problem—was being pursued. The undertaking, called the National Coal Policy Project, sought to achieve *some* consensus and cooperation between two long-term antagonists: industry and environmentalists. . . . In the National Coal Policy Project, they came together in a cooperative effort to reach consensus, *provided guidance* for the resolution of national coal policy issues, and to articulate their differences in a useful way. At the end of the first stage [in February 1978], the project's leaders claimed 80 percent agreement. They urged a next phase to implement their recommendations, and in December 1978 they at least announced plans to study a specific series of issues.[6]

The author, Mel Horwitch, certainly was not using inflated language. He saw the NCPP as an attempt to provide "some consensus," and cites *Where We Agree* to indicate that a goal of the project was to "provide guidance." Yet he praised the NCPP warmly:

> The project was attacked by some environmentalists, by some members of industry, and even by some persons within the government, but what was really important was that *a potentially significant, new cooperative process was set in motion*. Such an effort *might begin* to

> replace the worn, debilitating rhetoric of conflict with sensible compromise and with a *successful resolution* of difficult issues.[7]

Note that here a reference to problems of implementation is avoided by an ambiguity of reference; it is not clear whether the author expected the new cooperative process to replace the "debilitating rhetoric of conflict" just within the institution of mediation, or whether the reference is to the whole political system. If the entire political process had been referred to explicitly, Horwitch would have had to indicate how such "successful resolution" could be achieved beyond the bounds of the conference group.

The author continued:

> The 1977–78 coal strike and the National Energy Project offer two radically different examples of what coal's future might be. On the one hand, the United States can have a coal industry that suffers from internal conflict, labor strife, and environmental contention. . . . On the other hand, the United States can go down the road marked out by the National Coal Policy Project, finding methods to cope with coal's short-term difficulties and immediate conflicts.[8]

On close inspection, this idea is misleading. No one expected that the coal policy area would continually be wracked by conflicts as serious as those of the 1977–1978 coal strike. Actually the appropriate institutions for comparison to the NCPP are the normal institutions of decision making in American politics: the congressional process, environmental litigation, and so forth. On the other hand, Horwitch had a roughly accurate idea of the limited scope of the NCPP so far. He knew what the NCPP was, but he did not know how to think about it. So he concluded with a statement that the "calm and reasonable approach" of the NCPP is preferable to the 1977–1978 coal strike. In addition, he sloughed off the question of implementation.

The NCPP was praised in an important study, *Energy and America's Future*, conducted by Resources for the Future and published by Johns Hopkins Press in August 1979. The RFF writers admitted two of the problems stated by the NCPP's critics—that conflict resolution in the NCPP dealt with only some of the issues in the coal area, and that the NCPP could not enforce its agreements. They seemed to be of two minds in evaluating the NCPP. Perhaps their conclusion can be summarized: "It's not much, but it tries to do something that is essential to the future welfare of the country."

The RFF study stated that America's energy policy "must explicitly combine three broad goals," those of sufficient supply, conservation, and protection of environment and human health (p. 9). The authors were confident that such a strategy could be worked out, but that it was "most important" that there "be a will to seek a consensus" that is needed to realize these three broad, related goals.

> How might we go about achieving such a consensus? The National Coal Policy Project is perhaps the foremost example to date of an attempt to reconcile conflicting attitudes toward environmental protection, resource conservation, and energy supply. It could be a prototype of what will be required if such reconciliation is to be achieved on the full range of energy policy issues. An analysis of what this disparate group agreed upon, and what it *failed* to agree upon, indicates both its promise and its *limitations*.[9]

The RFF authors were certainly guarded in their assessment of the actual accomplishments of "the foremost example to date of an attempt to reconcile conflicting attitudes" in the energy area. And they explicitly recognized a basic problem in the implementation of *Where We Agree*: that corporations and public-interest groups have refused to endorse the actions of their personnel who participated in the NCPP:

> If their statements of common ground leave a good many issues still unresolved, we must admit at the same time that they are both broader and more specific than the areas of agreement the United States has staked out so far in connection with the overall energy picture. The National Coal Policy Project has not been widely accepted as a "breakthrough," even with respect to coal. The participants have yet to convince the divergent groups from which they were drawn that the consensus reached is worth supporting. Nevertheless it is a beginning.[10]

Saying that the NCPP has been more successful in finding "areas of agreement" than "the United States has staked out so far in connection with the overall energy picture" is obscure praise. Upon consideration, if "the United States" "staked out" an area of agreement, would we know this if we saw it? Is there currently an agreement to devote significant resources to solar energy, for example?

Later, however, the RFF authors did offer a rationale for their view that "the start made by the Coal Policy Project is significant and promising" as a step to increasing consensus on an overall energy policy.[11] In referring to how conflict over energy issues might be reduced among groups, the authors (mostly economists) applied a concept that seems to be a modification of Pareto optimality. Interest groups in negotiation are urged to (1) accept one another's minimum requirements; (2) find a strategy which meets these requirements; (3) realize that this strategy is probably no one's first choice; (4) realize that such a policy is better than no agreement at all, which would contribute to a national disaster in the energy sphere. In the words of the authors: "Detente in the multifaceted conflict over energy policy might be achieved through an approach that meets the *minimum* requirements for acceptability by all important groups, and thus creates an overall energy strategy which, while not the preferred path for *any* of them, is judged superior to some plausible outcomes by all of them."[12] The authors adumbrated this Paretian conception with another idea: parties should recognize which of them have "primacy"—the legitimate right to veto a solution—in a negotiation. In many situations, then, conflict resolution requires that the parties who gain compensate the parties having primacy.

The RFF authors concluded that the NCPP was an example of their theory of negotiation, and this was the basis for their assessment of the project as "significant and promising" even though it was not a "breakthrough" in the coal policy area. As the authors put it: "The potential for success from recognizing the legitimacy of each party's minimum requirements and then seeking compromise on all else is shown by the National Coal Policy Project."[13]

Such a treatment of negotiation is obviously not very complete, but the intuition of the panel is correct: their theory does provide insights into the workings of the NCPP. Thus, most persons who consented to be part of the NCPP with its "rule of reason" had de facto agreed to respect the minimum requirements of the other side: the need to make a profit, the legitimacy of the environmentalist groups as representatives of environmental interests, and so forth. And the negotiation theory of the RFF showed something interesting about the NCPP. In the project, "primacy" was given to the interests of the public as a whole as defined by the concepts of economics. For example, environmentalists agreed that it is inefficient to devote more money to scrubbing emissions from old coal-fired power plants than to building new plants with good scrubbing systems. On the other hand, industrialists deferred to the public's "primacy" in the need for internalizing the external costs of coal mining and of burning coal. The NCPP's participants also believed that even though no one was getting his most preferred option, both environmentalists and industrialists were better off compromising with one another than ignoring one another, which would only increase the power of government as the arbiter of conflicting interests. In summary, the RFF's justification for its praise of the NCPP had a rationale: a theory of bargaining. Clearly, however, the RFF analysis was incomplete, even in its own terms, for it never discusses how the NCPP—or similar conflict-resolving bodies—can significantly affect the ongoing political process.

OTHER PUBLIC RELATIONS

The NCPP's staff at CSIS conducted a public relations effort of medium scope during the years of 1978 and 1979. Twenty-three meetings were held, mostly with officials of agencies of the U.S. government, to brief those in attendance about the NCPP. Nine seminars were held at universities around the country to more formally present the work of the NCPP. Seventeen speeches were given at conferences and meetings of associations. Seven "formal submissions" were made to agencies of the federal government commenting on proposed regulations or recommendations for possible future regulations. Two congressional hearings were held on the project (see below). One of the briefings for members of Congress at CSIS was particularly successful, because Representative John Dingell (D-Mich.) became interested in the NCPP's concept and decided to hold a congressional hearing on the idea in April 1978.

Some impressions of the effects of the project's public relations work are in order. After a few years passed, the Washington environmentalist lobbying community tended to dismiss the project as a failure, because its recommendations seemed to have little effect on policy. Quite a number of environmentalists, on the other hand, remembered the project's own success in reaching agreements and were stimulated to consider whether some modification of the NCPP's approach might successfully affect public policy in some different situation (see the next chapter on regulatory negotiation). Such environmentalists tended to be associated with more "middle of the road" lobbies and organizations, such as the National Wildlife Federation or the Conservation Foundation. The reaction of those in the coal business is difficult to assess, as I have fewer contacts with this sector. Many leaders in the coal industry knew little about the NCPP; those that did perhaps also dismissed its impact as small. But my own experience is that the more business executives learned about the NCPP, the more interested in it they tended to be. One case in point: a conference on applications of the project to the coal industry was organized by a private consulting firm in November 1979.[14] At least 100 business executives attended, even though there was a registration fee of $350. The conference prospectus, which included the summary of *Where We Agree*, stirred up significant interest among executives in the coal and electric utilities industries, indicating a widespread, latent positive image of the project among this group.

As noted, the project received favorable and widespread attention in the press during the first two months after the publication of *Where We Agree*, but afterwards was not generally considered to be newsworthy. Academics, on the other hand, have consistently regarded the NCPP with favorable interest and, upon hearing about the project for the first time, often show considerable curiosity. Lawyers and law professors interested in experimenting with modifications of the normal adversarial process, a subject known as "alternative forms of dispute resolution," usually show an awareness and a respect for the NCPP. By the mid-1980s, a national network of experimenters and researchers in the field of alternative dispute resolution had formed who remember the NCPP as a pioneering effort, perhaps the most elaborate conducted in the field, an effort which did not come to immediate fruition, but did stimulate others to develop the idea of regulatory negotiation.[15]

The NCPP was somewhat successful at first in its public relations activities. Accordingly, its failure in lobbying is all the more surprising. The main cause of this failure was the lack of group endorsements for *Where We Agree*, including the nonendorsement of organizations having officers participating in the NCPP.

THE DENIAL OF GROUP ENDORSEMENTS

The lack of group endorsements can be understood in a straightforward fashion. Group politics often relies on consensus and minority veto. Environmentalist organizations and trade associations like the National Coal Association do not ordinarily take actions to which a sizable minority of their group objects to, and the NCPP was sufficiently controversial that minorities of environmentalists and coal company executives objected to it. This difficulty of the minority veto was compounded by a second phenomenon. Interest groups within a sector often lobby in coalitions. Environmentalist groups form lobbying coalitions; coal companies do the same. Consequently, if a proposal is seen to be of secondary importance, rather strong objection to it by two or three groups in a sector can be enough to kill the proposal, because the other groups do not wish to alienate their political coalition partners.[16] This need to placate allies was a major reason why the NCPP did not get group endorsements.

I chronicle the environmentalist objections to the NCPP at greater length than the business objections—not because I perceive environmentalists as inherently more opposed to the NCPP's style of decision making than business, but because, in this instance, environmentalists took the lead in opposing the platform of *Where We Agree*. If environmentalists had not performed this function, certain businesses would likely have filled the gap. More attention is paid to the environmentalists, furthermore, because environmental politics is fought more openly than business politics. Louise Dunlap of the EPC and Richard Ayres of the NRDC each met with the author for an hour each to explain their positions on the NCPP. Officials of the National Coal Association refused to speak to the author, on the other hand, apparently because of their embarrassment over an internal division, regarding the desirability of endorsing the NCPP's platform.

One difficulty for the NCPP was that its activities infringed upon the turf of established lobbyists. Each "turf" is well understood and well defined within the Washington lobbying community. (The term "turf" is actually used.) Groups specialize in lobbying for particular issues and bills. Other groups will then defer to the lead of a group that has developed an issue and previously expended considerable money and effort in lobbying for a position on an issue. In this case, Louise Dunlap and the EPC had occupied the position of environmentalist specialist on strip-mining issues, and Richard Ayres and the NRDC were seen as the environmentalist leaders on air-pollution control measures.

By seeking to develop a second, acceptable environmentalist position on strip-mining and on air pollution, then, the NCPP was infringing on the turf of the EPC and the NRDC. In Washington lobbying practice, this is not commonly done by groups on the same side of an issue. By entering someone's "issue territory," the new group is likely to be perceived as an opponent of the established lobbying groups. This happened to the NCPP.

The project also suffered from entering the turf of the National Coal Association and the individual lobbying efforts of separate coal companies. On strip-mining issues, the lead lobbyists were Harrison Loesch, of the Peabody Coal Company, and officials of the National Coal Association. Loesch, unlike Dunlap and Ayres, was enthusiastic about the NCPP and was one of the leaders of the strip-mining committee. His support perhaps canceled out the possibility of leading coal company lobbyists actively opposing the NCPP, as Dunlap and Ayres did. Still, Loesch, John Corcoran of the NCPP, and other coal executives supporting the NCPP were not sufficiently powerful to get endorsements from their own industry.

Frank Murray of Georgetown's CSIS, in charge of the implementation phase of the NCPP, encountered three audiences hostile to the project in the course of giving sixty speeches about the NCPP. One group was strip miners in southern Montana who thought the project represented eastern coal interests by not pushing for a great increase in the production of Great Plains coal; the other two opposed groups consisted of various coal company lobbyists in Washington who resented the project for legitimizing the SMCRA with some sectors of the coal industry and for developing a separate, second position on coal-related issues generally different from their own. In particular, circulation of *Where We Agree* meant that top coal-company bosses would read an analysis of coal-related issues different from the memos lobbyists had been sending to corporate headquarters. At the very least, this challenge to their authority irritated many lobbyists.

To be sure, some groups opposing the project were protecting their own organizational interests, but there were good, principled reasons to oppose the idea of the NCPP. Louise Dunlap and the EPC had fought for regulating mountaintop removal mining practices, for eliminating the dumping of excavated earth on top of rich farmlands, and for prohibiting strip-mining in the western plains. The NCPP proposed to modify prohibitions in these three areas. One cannot be surprised that Dunlap was critical of a group that proposed to modify her own political successes. Similarly, the project suggested modifications of some of the positions taken by Richard Ayres and the NRDC. For instance, Ayres strongly opposed the use of tall stacks to meet air pollution regulations. The NCPP, on the other hand, would have temporarily legitimated the use of some tall stacks on older electric power plants if the resulting savings were applied to the purchase of antipollution equipment for newly built plants. Finally, though it is difficult to describe corporate resistance to the passage of strip-mining controls as a matter of principle, we might note that a sincere argument existed that higher costs of production under the new regulations would mean fewer jobs in a chronically depressed industry.

The chief adversary of the NCPP, Louise Dunlap, stated in an interview that one of her reasons for opposing the project was that Larry Moss, Mike McCloskey, and others "misrepresented" the NCPP as a "consensus" of the relevant interests in the coal area. Dunlap stated: "They had a perfect right to do it [form the project], if they called themselves a group of individuals who got together to work out a platform. But it was advertised as more than that." Ms. Dunlap said her impression was that the leaders of the NCPP believed they were making better policy proposals than those adopted by Congress. "Who were Larry Moss and the National Coal Policy Project to set themselves up as the decision makers?" she asked. Dunlap indicated she was particularly angered by Mike McCloskey's statement in the 1978 article in *Fortune* about the NCPP:

> "I'm now persuaded that in many cases the possibilities for resolving policy issues are better in this kind of setting than they are in public legislative bodies," [McCloskey] says. "There's not the same overlay of extraneous issues—the opportunities for public advancement, the competition for attention. When a topic gets that extra political twist on it, it often gets simplified and horribly mangled. Confrontation politics leads to sweeping generalizations rather than fine distinctions and both sides conceal their ultimate aims."[17]

McCloskey's statement is an example of what might be called an "unbounded view" that the project was something like a legislature, that the

NCPP is a conclave of representatives of parties interested in coal, and that this conclave makes decisions that are binding upon these "representatives" and their organizations.

By 1979 no one was taking the unbounded view in describing the NCPP. At this time, and earlier in *Where We Agree*, leaders of the project emphasized that it was composed of persons acting as individuals, not as representatives of organizations. For instance, the preface to *Where We Agree* states (in all capitals): "The following individuals formally participated in the National Coal Policy Project. They took part as individuals rather than as representatives of their various organizations. Their affiliations are given for the purposes of identification and to show the broad range of experience and interests represented in the project."[18] Again, in the introductory section of *Where We Agree*, under the section heading (in boldface type) "Individuals, Not Representatives," it is stated:

> The participants in the project took part as individuals. Although they were selected in part because of their leadership roles in environmental and industrial organizations, they do not purport to speak either for their organizations or for the environmental and industrial communities at large. The issues are too complex and controversial for either side to speak with a single voice.[19]

Overselling the project might have been efficacious during the initial stages of organization in getting funding and participation. Jerry Decker, Larry Moss, and the others probably took a limited view but thought that many environmental organizations and coal companies would give clout to the proposals by lobbying for them. The fact that no lobbying occurred presaged an important change in perception of the NCPP which occurred between 1976 and 1979. Nevertheless, challenging the credentials of the NCPP to represent the interests of the public is certainly an important criticism, an example of Dunlap's going beyond a simple defense of the organizational interests of the EPC.

In any case, Ms. Dunlap staged an extraordinary political maneuver that dealt the death blow to gaining endorsements for *Where We Agree* from environmentalist lobbies. Dunlap persuaded the Sierra Club Governing Board to order the club's executive secretary, Mike McCloskey, not to participate in the NCPP. McCloskey had spent at least fifteen working days during the year of 1977 developing the report of the mining committee. The work occupied a major portion of his time, even though the Sierra Club was not paying his salary for this effort. At least one board member demanded that he account for his activities on the project, especially as the Sierra Club was becoming identified with the NCPP, due to the prominence of Moss, McCloskey, and Greg Thomas (a well known club lobbyist) in the institution.

A club board member, suspicious of the project, invited Dunlap to speak at the board's meeting in January 1978, when McCloskey's activity in the NCPP was discussed. At this meeting, a debate arose between Larry Moss (who spoke in favor of the project) and Louise Dunlap: Ms. Dunlap won the argument as far as the board members were concerned. They then voted to forbid McCloskey to participate further in the project. The EPC head expressed her usual arguments: (1) the project was not representative; (2) the project was a sell-out to the coal companies; (3) existing legislation was better than that advocated by the NCPP; (4) the idea that the project was a better decision-making institution than Congress or the courts was "arrogance"; (5) environmentalist groups had scarce resources and should not waste them by having their leaders work on worthless efforts like the NCPP.

However, Frank Murray and other NCPP leaders had the impression that a majority of the board might have supported them in different circumstances. But probably a simple majority of the club's board would not be enough to allow McCloskey to work on the project. A minority had blocked the passage of a resolution criticizing nuclear power by the club's board until 1974. A different minority could be expected to block endorsement of *Where We Agree*.

Dunlap's defeat of the project in "the battle of the Sierra Club" ensured lack of endorsement by other environmentalist groups. Why would the Audubon Society, the Environmental Defense Fund (EDF), or the National Lung Association endorse substantial portions of *Where We Agree* when it was well known that the EPC, NRDC, and Sierra Club had refused to do so? With the backing of the Sierra Club, which in the mind of the general public was *the* environmentalist lobby, it would have been possible to criticize the EPC and NRDC as expressing environmentalist militancy in opposing the NCPP. Endorsement by two or three other environmentalist groups—presuming it is possible to overcome the reluctance to override minority opposition on a board—would have placed the project in a good position to get endorsement from the coal association.

The resultant "bipartisan" organization support would have seemed most impressive to members of Congress and to journalists.

But the project's tactical defeat meant that no leading environmentalist group would go out on a limb to endorse *Where We Agree*. The NCPP report was not seen as sufficiently important by members of environmentalist boards to override the minority opposition that would surely occur. Why would a member of the EDF governing board, say, start a fight within his organization to get support for *Where We Agree*? There are plenty of worthy causes.

Another factor explains the lack of endorsement by organizations—both environmentalist and industrialist. Virtually all political groups want to maintain good relationships with their normal coalition partners in lobbying. Indeed, lobbying in Washington is lobbying by coalition, for the most part.[20] On important, general legislation, such as bills to control mining or amendments to the clean air legislation—there are scores of groups on each (or several) sides of the political debate. It is not skillful politics to have, say, seventy-five political groups each sending a lobbyist down to the Hill without some effort at coordination: the average member of Congress is interested in talking to perhaps five lobbyists on a given matter, but does not want to talk to a hundred of them.

Rivalry and some rancor usually exists among political groups in some area of lobbying, such as environmentalism or coal mining. However, if the lobbyists within some sector are to enhance their overall effectiveness, they must work together in tactical struggles within Congress. And, in my judgment, the leading environmentalist lobbies in Washington have succeeded in so working together. There are few (if any) instances in which differing environmentalist lobbies have embroiled themselves in a major public battle before the eyes of Washington. Environmentalists—whatever their internal rivalries—see themselves as all promoting a just cause in the face of the overwhelming odds posed by the enormous resources of corporate America.

Therefore, when three environmentalist organizations are opposed to an idea, it is nearly impossible to get other such organizations to endorse it. If the idea is not seen as central to immediate political struggles—as was the case with the NCPP—then an environmentalist organization will not want to offend one of its frequent coalition partners.

Probably *Where We Agree* would have run into a similar problem among industrial lobbyists, if the issue had been drawn. For example, endorsement of the mining committee report was a recognition of the need for government intervention in the conduct of surface mining. Industrial groups outside of the coal area might have resented an endorsement of government regulation by the National Coal Association (NCA), even though the overall tenor of *Where We Agree* is to rely upon the market mechanism. The NCA was looking forward to the time when it could successfully press for amendments to the 1977 strip-mining act. Of course the NCA also lobbies for quite a number of other goals and can use all the help it can get. The NCA was disastrously defeated in 1983 by the railroad lobby over the question of using federal right of domain to further the construction of coal slurry pipelines.[21]

In regard to its own internal organizational dynamics, the NCA would have resisted an endorsement of *Where We Agree* for the same reasons that some environmentalist organizations would have resisted. As noted, Frank Murray observed that a number of professional coal-company lobbyists were opposed to the project. Some of these people would have resisted an endorsement of the NCPP by NCA's board. This was why, in Murray's opinion, the NCA did not endorse *Where We Agree*. Murray believed that the majority of staff and NCA board members were, as individuals, favorable to the project's platform.

Murray's view is probably correct. Of course, in a crisis, the leadership of an interest group will bear down on a minority to support some principle that is deemed important. But in 1978, the principles of *Where We Agree* would not have been sufficiently important to the leadership of the NCA to offend its lobbyist members who felt threatened by the NCPP. Apparently the NCA had some problem with the NCPP, because its officials would not discuss it with me. Mr. Glen Schleede, described as "the policy think person" of the NCA, refused to speak to me, even after receiving an initial letter, followed by four phone calls, appropriately spaced. Mr. Joseph Mullan, who had been the NCA observer at several meetings of the project, did talk to me. He told me he definitely thought that the NCPP was a good idea because it had provided a basis for communication between the coal industry and environmentalists. "Anytime that happens, it is a good idea." But Mullan pleaded that he was an engineer, who did not understand politics, and thus would not talk about the project any further. Such behavior on the part of the NCA officials coincides with the observations of Murray that most of them supported the conclusions of the NCPP, but did not want to say so in public for fear of angering some lobbyists for the coal industry.

However, it must be noted that the NCPP took an unaggressive stance to the coal association. When asked why the NCA did not endorse the proposals of the mining task force, John Corcoran, industry leader of that committee, stated succinctly: "No one ever asked them to." Corcoran stated that "I have heard no negative word" about the NCPP from persons in the coal-mining industry and that he had "heard many positive comments about the process."[22] These remarks corroborate the impression that most of the thirty-five or so members of the NCA board would have endorsed the mining committee's proposals, if they had been forced to make a decision. But an endorsement of the mining committee's report by the coal industry would give the wrong impression if the report were not endorsed by several leading environmentalist organizations, for the mining report might then be regarded as coal company propaganda. *No* endorsement from the NCA was better, from the standpoint of the NCPP, than an NCA endorsement without accompanying environmentalist endorsements.

We now can see why there was almost no incentive for individual participants in the NCPP to persuade their own companies or lobbies to endorse *Where We Agree*. A few initial soundings must have been enough to divert such individuals into other, more popular, activity within their organization. Only Jerry Decker would have risked getting fired for pushing an endorsement of the project, but during the course of the meetings preceding *Where We Agree*, Decker left Dow Chemical for the Kaiser Aluminum Company.

The minority veto barrier to group endorsements of the NCPP is a more general phenomenon among contemporary interest groups. Apparently most political lobbies avoid endorsing a measure when this action is expected to offend a substantial fraction of the group's members (one-fifth is the order of magnitude I have in mind).

Case materials supporting these generalizations can be found in my book, *Public Interest Lobbies: Decision Making on Energy*.[23] This study seeks to explain the energy stands of seven "public interest" groups by examining the political and organizational factors that constrained them. A negative example will introduce the point. It is well known that the American Civil Liberties Union lost 25 percent of its members after it defended the right of Nazis to march down the main street of Skokie, Illinois, a predominately Jewish suburb of Chicago.[24] This incident is widely remembered because it is so exceptional, the sort of thing that the board and staff of political groups try to avoid. Who wants to lose one-fifth of one's budget? Who wants to spend one-fifth of one's time answering angry telephone calls and letters?

In field work at the Common Cause national headquarters, I soon realized that the leaders of this organization pull back from a policy that would be opposed by 20–25 percent of its members.[25] For example, although a majority of Common Cause members were critical of the value of nuclear power plants, a minority supported the need for such plants. Accordingly, after an internal conflict, Common Cause refused to endorse an antinuclear initiative in the state of California in 1976. The leaders of the organization prefer to concentrate their lobbying on measures that have virtually unanimous support among the Common Cause membership—such as lobbying for ethics codes for members of Congress. My observation of other public interest lobbies confirms that such lobbies avoid offending a substantial fraction of members. For instance, the Consumer Federation of America's Energy Task Force did not take stands on nuclear or coal issues, because some of its members were rural electricity cooperatives that used power created in nuclear or coal plants. The Sierra Club, which has a more diverse supporting coalition than the new-style environmentalist lobbies, was quite late in opposing nuclear power (1974).[26]

Because ordinarily only a minority (about one-fifth, say) of a political group's board is needed to block the adoption of compromise proposals such as those on the project, an opponent of the compromises holds a very favorable strategic position to block an endorsement.

THE NCPP AND CONGRESS

The relationship of the NCPP to the U.S. Congress may teach us the vagaries of the political process. The NCPP had a little impact on the Congress, but not much. A House Commerce subcommittee held a hearing on the NCPP, at which the project's leaders testified, presenting the conclusions of *Where We Agree*. A similar hearing was held by a Senate Government Operations subcommittee.[27] (See the next chapter.) These hearings, however, were not related to specific bills introducing legislation taken from the NCPP's report. Though such legislation was introduced by a member of the House, Don Pease (D-Ohio), no subcommittee hearing was ever held on Pease's bills. Leading House members—Al Ullman of Oregon, the chair of the Ways and Means Committee; John Dingell (D-Mich.), influential on air pollution legislation; Don Fuqua of Florida, the chair of the House Science and Technology Committee; and Richard Ottinger (D-N.Y.), a leading liberal spokesperson—were all aware of the NCPP, were in contact with NCPP staff, and seemed favorably impressed by the project.[28] Nevertheless, this favorable impression did not lead to much. The NCPP had little impact on the Senate, although it did impress newly elected Democrat Carl Levin of Michigan as a novel mode of dispute resolution which augured well for regulatory negotiation and similar experiments in dispute resolution.[29] Staff of the NCPP did meet with staff of Senate Majority Leader, Robert Byrd of West Virginia, and coal state senators Wendell Ford, Democrat of Kentucky, and Charles Percy, Republican of Illinois. But these meetings did not seem to have much effect.[30]

The NCPP's leadership directed its energies to getting agreement among project participants, and in this they were successful. Of course the leadership did not expect the interest-group vetoing effect, which stymied the project's impact. If all the environmentalist lobbies in the NCPP, together with the coal companies and the National Coal Association, had endorsed *Where We Agree* and had followed up such an endorsement with vigorous lobbying, most members of Congress probably would have been impressed. In this scenario, the project's own lobbying would have taken second place to that of the environmentalists and the coal association. As we have seen, however, even a significant compromise platform is likely to be rejected by participating interest groups, leaving the joint effort to die.

A better institutional design would have included congressional staff in the process, either as negotiators, mediators, or expert consultants. At best, such congressional staff would be associated with committees having leading decision-making roles in regard to the NCPP's proposals: the Senate Energy and Natural Resources Committee, the House Interior and its commerce committee, and so forth. Congressional staff participation might have committed a few leading members of Congress to endorse and push for NCPP proposals.[31] Members of Congress often contact interest groups and urge them to support pending legislation.[32] With staff participation, there would have been some chance that members of Congress could influence recalcitrant interest groups to back some of the NCPP's compromise proposals. But congressional staff participation was not considered by the leaders of the NCPP, although staff were invited to witness the meetings of the NCPP's plenary group. Congressional staff participation would have violated the voluntaristic self-conception of the project, whose leaders defined the project as posing an alternative mode of defining policy, separate from the government.

Another major factor retarding congressional interest in *Where We Agree* is that it was published during an unfavorable phase of a legislative cycle. In the summer of 1977, after years of major legislative battles, Congress passed major strip-mining and clean-air legislation. In 1978, then, Congress was not in a mood to reopen the stormy controversies regarding strip-mining and clean-air legislation. Indeed, since 1977 Congress has not passed major amendments to SMCRA of the scope advocated by the NCPP, and it waited until 1990 before passing another important set of amendments to the clean air acts.

There would have been much more interest in *Where We Agree*, even without endorsement by interest groups, if President Ford had been reelected in 1976. In this case, the strip-mining issue probably would not have been settled in 1977, because Ford so insistently vetoed strip-mining legislation backed by 60 percent of Congress. Clean air amendments might have passed in 1977 if Ford had been president, though the accompanying controversy would likely have been great, as the Ford administration would have been dealing with a Democratic Congress.

Accordingly, if Ford had been reelected, much more frustration would have arisen with the squabbling and the delays in the normal decision-making process. With some astute lobbying on the part of the NCPP, *Where We Agree* could have been used to bridge differences between congressional Democrats and a Ford administration. President Jimmy Carter, on the other hand, was in close accord with Democratic majorities in Congress on the issues of strip-mining and clean air amendments. Legislation passed rapidly and was signed by the president (see Chapter 2), so there was no need to use *Where We Agree* to bridge a political gap created by the separation of powers.

In his study of how issues reach the Washington agenda, John W. Kingdon found it useful to distinguish processes in which problems are posed for legislators, processes in which solutions are offered (such as the platform of *Where We Agree*), and the political process, which at certain times offers a "window of opportunity" for legislation, when political entrepreneurs match a solution with a problem.[33] Unfortunately for the NCPP, the years 1978–1981 did not pose a favorable window of opportunity for passage of *Where We Agree*. In the strip-mining policy area, during the 96th Congress (1979–1980) the "problem" for the coal industry was to decentralize implementation of the SMCRA away from the federal government to state government, for such states as Kentucky and Wyoming could be expected to enforce the SMCRA loosely, given the power.[34] Environmentalists, on the other hand, had wanted to maintain centralized enforcement of the SMCRA in the Office of Surface Mining of the Department of the Interior.[35] The Senate passed a bill in 1979 congruent with the coal industry's position, but the House took no action.[36] In this situation, environmentalists had no incentive to support the NCPP's strip-mining position, which called for flexible administration of the SMCRA, because it would have weakened their argument against the coal lobby's call for implementation by the states. In 1979–1980, the project's strip-mining platform was no answer to coal policy problems as defined by the lobbyist participants.

Similarly, the years 1979–1980 offered no window of opportunity for the project's air-pollution platform. Clean air issues are notable for their complexity and for producing a great mobilization of interest groups. After fighting over air pollution issues in 1976–1977 (see Chapter 2), Congress had exhausted its interest in proposed Clean Air Act amendments. The people I interviewed for this study and my daily reading of the *Washington Post* at that time indicated clearly that Congress wanted to digest the Clean Air Amendments of 1977 and consider the issue a few years hence, presumably in 1982 when the law needed reauthorization. In 1979, few members of Congress worried about the problem of possible shortages of electricity in the early 1990s, and so the project's solutions in this area, such as facilitating siting decisions for new power plants, incited little interest.

Still, the NCPP might have had more impact on Congress, maybe getting a floor vote on one of its ideas, if the project had been poised to lobby Congress in March 1978. Instead, though the project's staff was active in approaching journalists and academics, active lobbying of Congress was infrequent, largely confined to a seminar for members of Congress and congressional staff, which did persuade Representative Dingell to hold a subcommittee hearing on the project. But not much else was done to lobby Congress in the fifteen months before the summer of 1979. One problem was that employees of Georgetown University were not permitted to lobby Congress, and the project's staff and staff director Frank Murray were officially employees of Georgetown. Although Murray believed that holding a seminar for members of Congress was not lobbying, he did not feel he could approach members of Congress or their staff on behalf of the NCPP's platform.

Legally, however, to approach a member of Congress to discuss an issue is not "lobbying," unless one tries to sway the member's opinion on some specific bill.[37] Accordingly, leaders of the NCPP's plenary group persuaded Murray to hire a congressional liaison, who worked for the project from the summer of 1979 to the summer of 1980. The liaison, Ralph Nurnberger, a former Senate staffer, understood the ways of Capitol Hill and proceeded to distribute copies of *Where We Agree* to all members of Congress, to organize meetings with congressional staff, and occasionally to meet with a member. However, by the fall of 1979, the political window of opportunity for the NCPP had almost closed, and Nurnberger could do little except precipitate the introduction of legislation by Representative Don Pease.

In summary, the NCPP's impact on Congress would have been enhanced if Congressional staff had played an active role in the project's negotiations. Unfortunately for the NCPP, the legislative cycle limited its possible impact on Congress, something to be remembered in future experiments in cooperative pluralism.

REGULATORY SUCCESSES

The failure to translate most of *Where We Agree* into public policy seems initially to pose grave questions about the viability of cooperative pluralism as public policy. Further analysis presents a more optimistic picture, however, for a limited portion of the NCPP program *was* enacted.

One major success of the NCPP was that it influenced the writing of regulations by the Federal Energy Regulatory Commission (FERC, a division of the Department of Energy) on issues related to cogeneration of electric power. These regulations were issued by FERC in February and March 1980 to implement the federal statutes known as the Public Utility

Regulatory Policies Act, passed in December 1978. One of the more successful areas of agreement before the publication of *Where We Agree* was cogeneration policy. This issue was a natural for the NCPP. Cogeneration seemed to be an ideal means to produce more electric power without increasing the consumption of coal or oil. Environmentalists and business could agree on the need to further cogeneration and then trade off positions on two basic issues: in 1977, major obstacles to increasing the use of cogeneration included federal laws defining a "public utility," and federal and state laws specifying the obligations of an electricity-generating public utility. A factory wanting to cut costs through cogeneration could find that burdensome regulation under public utility laws cost more than cogeneration would save. From the point of view of business, one obstacle to cogeneration was opposition from electric utilities in areas of relatively static demand; another was a continuing problem of an excess generating capacity. Such utilities often opposed an increase in cogeneration and might charge higher rates to a cogenerating factory that might need to buy additional power from the electric company. Here environmentalists could criticize excessive government regulation, and the business side in the NCPP criticized behavior of a small segment of the overall business community which hindered the national effort to conserve electricity.[38] Hence, in the NCPP, environmentalists and business could readily strike a deal on cogeneration.

In this policy area, opinion in the NCPP resembled opinion in the U.S. Congress. Although the project's negotiators were dormant during 1978 after the publication of *Where We Agree*, Congress passed the Public Utility Regulatory Practices Act, generally referred to as "PURPA," which contained sections designed to deregulate cogenerators from the restrictions imposed by definitions of "public utility." One of the three task forces established during the second phase of the project's negotiation process was one on cogeneration policy. The Cogeneration Task Force met nine times between January 1979 and February 1980, and FERC actually issued its cogeneration regulations before the task force completed its report to the plenum of the project. But by 1979 the leaders of the NCPP were much more concerned about the need to lobby for the project, since 1978 had been a failure in this respect. Cogeneration policy seemed to be one area in which the project could actually affect policy, particularly as the FERC's staff were interested in what the NCPP's negotiators might recommend. Accordingly, the Cogeneration Task Force and the FERC's regulation writers conducted two joint meetings during 1979, and good communication was established between the two groups. Consequently, the federal regulations concerning cogeneration under the PURPA bear the imprint of the project, even though the task force did not draw up its final report until after the regulations were issued.

The Cogeneration Task Force described the essential regulatory issue under the cogeneration section of the PURPA as follows: "Congress enacted provisions . . . requiring the Federal Energy Regulatory Commission (FERC) to prescribe rules pursuant to which utilities must offer to buy and sell power from and to cogenerators at rates that are just and reasonable and that do not discriminate against cogenerators."[39] The exact provisions adopted by the FERC at the behest of the project are technical and meaningful only to the specialist in utility policy.[40] But to give some flavor of the issues involved, most of which have to do with the conduct of electric utilities towards cogenerators, I note two of the ideas that were written into the PURPA's specifications. First, utilities must offer a backup service to cogenerators in case the facility breaks down. Obviously, a utility might refuse such service to a factory to discourage cogeneration. Utilities were forbidden to charge their highest rates, such as peak-load rates, for the service. Second, if a cogenerator produces more power than it needs, what price should the utility pay for such power? Obviously, there is only one buyer, the local utility. The NCPP-FERC position was that "the cogenerator must be paid the utility's full 'avoided costs.'" These are the "costs to the utility of electric energy or capacity, or both, which, but for the purchase from the cogenerator, the utility would generate itself or purchase from another source."[41] The NCPP *Final Report* describes other overlaps between the NCPP and the FERC in regulating the price of power sold by the cogenerator:

> [1. The principle of avoided costs.]
> 2. A new cogenerator will have the right to sell all or part of its output to a utility at the utility's avoided cost even if the utility simultaneously sells energy to the cogenerator at its retail rate.
> 3. Each utility will be required to publish standardized tariffs setting forth the rates the utility will pay for power from facilities with a design of 100 Kw or less.
> 4. Utilities and industries will be free to negotiate rates that vary from the published rate tariffs or from the principles set forth above.

> If the parties fail to reach a negotiated agreement, however, the cogenerator may always avail itself of the FERC-ordered "avoided cost" rates.[42]

The project had a major impact upon these cogeneration regulations, and this was the NCPP's most important direct effect upon public policy, as opposed to such indirect effects as providing a pioneering model for negotiations among groups or improving the tone of communication in the coal policy area. Cogeneration itself is an important economic function that can save a considerable amount of energy,[43] though one should avoid extreme expectations of cogeneration becoming a basic tool for solving a future American energy shortage.

Success in the area of cogeneration accords well with the technical spirit of the project's proceedings. Indeed, after retiring from the NCPP and the Kaiser Aluminum Company, project founder Jerry Decker became a cogeneration consultant, establishing the Decker International Corporation for this purpose.

The NCPP also succeeded in the area of performance bonding requirements for operators of small strip mines under the strip-mining regulation law of 1977. One problem with the original regulations written by the Office of Surface Mining (OSM), a division of the Department of the Interior, was that the financial requirements for posting required performance bonds imposed hardships on operators of small mines. Under the law, mine owners are required to post bonds that they forfeit if they subsequently violate the reclamation or other provisions of the law. Presently, there are numerous small-scale strip mines, particularly in eastern Kentucky, West Virginia, and western portions of the state of Virginia. Such small mines each have only a few employees and are sometimes owned and worked solely by members of the same family. These mines are frequently unsafe and present special problems for the enforcement of federal mine safety regulations. Many small coal mines are economically marginal operations subject to abandonment or shutdowns during periods of lesser demand for coal. Enforcement of federal safety or reclamation regulations increases costs and is frequently met with an uncooperative attitude from owners and employees, who lose their jobs if the mine is shut down and who will have difficulty finding other jobs in the high-unemployment area of Central Appalachia.

Accordingly, it made sense to both environmentalist and business participants in the NCPP to advocate a relaxation of bonding requirements intended for larger coal companies to increase the incentive for the small operators to comply with the strip-mining law. This particular economic group had not been represented in the intense conflict between the environmentalists and the larger coal companies in the battles over the passage of SMCRA, the strip-mining control act.[44] Regulation writers for the OSM were not initially disposed to shed a tear over the fate of "mom-and-pop" strip-mining operators. Yet the case for modification of performance bonding requirements for this group was very strong. Leaders of the NCPP made formal presentations of suggested modifications in the rules for performance bonding to officials of the OSM and in March 1980 that agency released a set of modifications that reflected the NCPP's perspective.

An undesirable environmental consequence of strip-mining has been the abandonment of mines without any provision for reclamation and without much prospect for the government or legal authorities to identify the party responsible for continuing environmental degradation from the abandoned mine. Piles of removed earth would be a current landslide hazard; exposed rocks might leach to create acid run-off, with effects similar to acid rain; the hydrology of an area might be altered, and so on. Abandoned strip mines (and underground mines, also) constitute a long-term environmental hazard, primarily a hydrological one, perhaps, with effects on flooding and water quality and availability. The NCPP concluded that such hydrological effects of abandoned mines should be viewed in terms of water-basin units, similar to the analysis of air pollution by units known as air basins. The different impacts of abandoned mines on water could be related to one another scientifically. Like air pollution, environmental impacts on water transcend state boundaries, but there is a tendency to parcel out problems by the man-made boundaries of the state. The NCPP selected a specific water basin, the Tug Fork River Basin of West Virginia, Virginia, and Kentucky, that is subject to severe flooding problems and suggested that abandoned mine reclamation there be planned as a unit. The OSM accepted this suggestion and initiated such a program in 1980.

The NCPP also suggested that a national inventory of abandoned mines be an early priority of the OSM, which started such inventories in several states in 1980–1981, though not clearly at the suggestion of the NCPP.

The leaders of the NCPP did show some potential for persuading members of Congress to take a favorable view of the project, even though the NCPP is not clearly linked to any legislation from the U.S. Congress. Representative John Dingell (D-Mich.) was favorably impressed by a NCPP presentation in January 1978 and accordingly held a day-long hearing on April 10, 1978, at which eighteen leaders of the project gave a presentation to the Subcommittee on Energy and Power, which Dingell chaired, of the House Commerce Committee. The presentation included the history, philosophy, and platform of the NCPP, plus a discussion with the five congressmen who attended. A well-known liberal congressman, Richard Ottinger (D-N.Y.) was quite impressed by the presentation and subsequently showed particular interest in the idea of dispute resolution by negotiating among groups. But another congressman, Doug Walgren (D-Pa.) was angered by the Transportation Task Force's endorsement of government fees for river barge traffic, although Walgren claimed to like the general tenor of the NCPP.

Dingell did not introduce legislation to implement *Where We Agree*. But putting a 347-page volume of the hearing transcript on the record was useful to the project, by providing a ready reference source of a generally positive character.[45]

Representative Don Pease (D-Ohio) did introduce several of the project's main recommendations in two bills; the set of two bills was introduced in the Ninety-sixth Congress (1980) and then reintroduced in the Ninety-seventh Congress (1981). One bill, H.R. 1430 of the Ninety-seventh Congress, was entitled: "To expedite the decisionmaking process with respect to the siting of new coal-fired power plants and to provide, where possible, that such plants be located in the general area where the energy is to be distributed." The second bill, H.R. 1431 of the Ninety-seventh Congress was entitled: "To amend the Clean Air Act to encourage owners of coal-fired powerplants to utilize new technologies for pollution control and to establish an emissions charges and rebate plan." Neither bill was seriously considered; no hearing was held on either. The political time was not ripe for considering the bills. The Reagan-Watt confrontational approach on environmental issues ended tendencies of environmentalists to negotiate on siting questions. Similarly, the Reagan Administration's unsuccessful effort to rewrite existing air pollution control legislation created a legislative confrontation that left no room to consider measures advocated by the NCPP.

In sum, the NCPP gained some success that reflected the consensual, technical emphasis of the participants: it had an impact on cogeneration and reclamation-bonding regulations. But no legislation bore the project's imprint.

CONCLUSION

At this point, the prospect for getting public support for a mediation conference like the NCPP is indeed bleak. We might note that two different types of groups are likely to create political roadblocks to endorsements. As many would expect, groups that are militant or highly committed to a particular goal will object to compromising. And many such interest groups are now active in American politics; various social movements of both the left and right have left a legacy of lobby groups. The opposite type of interest group may also create problems in gaining endorsements. Business lobbyists and trade association officials will sometimes oppose groups that take a different approach to Washington lobbying and public policies from the one relayed by the "government relations" official to his corporate bosses. Both ideology and self-interest can combine to stymie the endorsement process of a mediation conference's proposals. Furthermore, if the proponents of the compromise positions overcome the minority veto of those on one side of an issue, they are still subject to a minority veto from the other side. Endorsements are not much good if they all come from groups on one side of an issue, because the report of the mediation conference appears to be a partisan document.

Yet the picture of the implementation of the NCPP does have some bright spots. Standard lobbying did succeed in translating a few proposals into public policy. The example of the NCPP furthered experimentation with regulatory negotiation, a sort of NCPP in microcosm, described in the next chapter. And the times may change, bringing an era in which politicians and public opinion would give high priority to experiments in reconciling political conflicts.

8

The NCPP and Regulatory Negotiation

Even though the NCPP failed to get most of its proposals adopted by policy-makers, its record encouraged later experimentation with negotiating differences among interest groups. Such experiments are known as "negotiated rule-making" or "regulatory negotiation."[1] But this concept differs from the NCPP in two important ways. First, regulatory negotiations include governmental officials. Second, regulatory negotiations focus on a limited set of issues. Recent years have seen a growing interest in this concept.

THE ROLE OF THE STATE

As voluntarists, the NCPP's leadership praised the role of interest groups—including both environmentalists and business—in policy-making and downgraded the role of institutionalized bureaucratic administration, or the "state." The NCPP's leadership instructed the project's participants that environmentalists, profit-making business, and society as a whole would be better off if policy were made by groups discussing political issues rationally and amicably, rather than by hidebound bureaucrats, who tend to invent unneeded regulations and then to enforce them rigidly and irrationally.

As we have seen, however, the voluntarist ideology of the NCPP was not practical. Voluntarism did not consider the difficulties of establishing a grand coalition of opposing interest groups, gaining the adoption of *Where We Agree* in public policy through its inherent persuasiveness, bolstered by the impressive political power of the united interest groups.

One aspect of voluntarism was a belief in the necessity of eliminating government negotiators from the NCPP process. There was an obvious argument for including, say, officials of the EPA, the OSM, or the Department of Energy. These were the people who would eventually write the regulations that the NCPP was trying to affect. But the leadership circle believed that including government negotiators in a tripartite process would greatly complicate the process of reaching agreement, and that government officials would be very inflexible in the negotiating process, because they would need to clear any changes in negotiating position with superiors outside of the NCPP discussion process. The project leaders apparently believed that government officials would jump on the NCPP bandwagon, once agreements were reached and endorsed by the grand coalition of groups. These expectations seemed reasonable in 1976, but the lack of endorsement for *Where We Agree* in 1978 undermined the basis of voluntarist views.

Contemporary bargaining/negotiation and mediation theory in the social sciences clearly would predict the failure of the NCPP because of the absence of government officials among the negotiators. For instance, in *Resolving Environmental Disputes: A Decade of Experience* (1986), Gail Bingham distinguishes site-specific environmental disputes (e.g., whether a factory should be built on some site) from policy dialogues, in which negotiators make recommendations about a policy.[2] She discovered seventeen such dialogues. Three of these included a party with the authority to implement a decision, and in each of the three cases the recommendation was partially implemented. In the other fourteen cases, no party had the authority to implement the recommendation, and in seven instances, the agreement was implemented anyway, but in seven other instances (including the NCPP) the agreement was not implemented.[3] In terms of both site-specific and policy-recommending environmental negotiations, Bingham asserts: "The most significant, measurable factor in the likelihood of success in implementing agreements appears to be whether those with the authority to implement the decision participated directly in the process."[4]

Bingham's conclusion implies that normally it would be a good idea for government agencies with enforcement powers to participate in environmental negotiations. In discussing site-specific negotiations about development projects, Timothy Sullivan states: "The inclusion of *representatives of government agencies* in negotiations may prove critical to the success of failure of this review process."[5] Mediation theorists Susskind and Cruikshank state the issue nicely:

> After the agreement is ratified, the negotiating parties must find a way to link the ad hoc informal agreement they have fashioned to the formal decision-making processes of government. Up to this point, typically, the negotiating process has been kept "unofficial." (It may have been this very ad hoc quality, in fact, which persuaded some of the key players to participate.) An unofficial process had produced an informal result, which probably could not have been reached by other means; the challenge now is to formalize that result. It seems paradoxical: How can an informal agreement be formalized?[6]

The paradox of moving from the unofficial to the official is certainly descriptive of this history of the NCPP. The paradox is more readily resolved if government agencies participate in the informal processes.

Actually the project's leaders moved beyond their early voluntarism during the second phase of the NCPP in 1979. Provoked by the lack of progress in implementing the project's proposals, the NCPP's leaders emphasized drawing up the specific proposals regarding cogeneration and mine bonding. This process was accompanied by meetings with regulation writers and officials of the enforcing agency—the FERC or the OSM.

In short, government officials were close to becoming a part of the NCPP's negotiating process; the original conception of the project was changed to move it closer to a tripartite negotiating process. Project leaders moved away from their original voluntarism to bring officials of the state closer to the negotiating process. In terms of impact upon public policy, this was the most successful phase of the NCPP.

In its second phase, then, the project moved away from voluntarism and closer to what is now known as "regulatory negotiation," which is usually seen as a negotiation process including government officials and directed to agreement on a specific set of regulations, such as the cogeneration regulations, in contrast to a search for agreement on general issues of policy, the earlier NCPP process reflected in *Where We Agree.*

REGULATORY NEGOTIATION

Several years after the NCPP disbanded, it now appears that the main impact of the project has been to encourage experiments with regulatory negotiation. According to its chief proponent, Washington attorney Philip J. Harter: "Regulatory negotiation is a process by which representatives of the interests that would be substantially affected by a regulation meet together to develop the initial draft of the regulation through direct negotiations."[7] Harter and most proponents of regulatory negotiation now advise that representatives of the regulatory agency participate in this negotiation process.

Like the NCPP, regulatory negotiation involves representatives of interest groups negotiating and compromising on proposed government regulations. A major difference is that regulatory negotiation is tied to a specific set of regulations, usually issued by a single government agency. Regulatory negotiation has much more specific goals than the NCPP, which aimed to develop a platform to deal with most of the major issues in an entire industrial sector. Despite these differences, efforts at regulatory negotiation were partially inspired by the NCPP, because the project's success in gaining agreement on so many issues indicated greater possibilities of agreeing on particular sets of regulations than many might have thought.[8]

Regulatory negotiation processes are normally linked to the agency that is to issue the proposed regulations.Thus, the basic institutional design of regulatory negotiation avoids much of the NCPP's implementation failure.

Jerry Decker and Larry Moss were not the only Washingtonians in 1975 who conceived of experimental mediation conferences between public-interest lobbyists and corporate representatives. Sam Gusman of the Conservation Foundation, an environmentalist foundation specializing in research on conflict mediation, began in 1976 a series of small experimental meetings of environmentalists and business executives to discuss issues of mutual concern.[9] These experiments became known as "dialogue groups," and by 1979 had acquired a reputation similar to that of the NCPP—internally successful at reaching compromises, but without much impact on the EPA or on other government agencies.

Another such experiment was the so-called "offeror process" of the Consumer Product Safety Commission (CPSC). In this process, the

CPSC would delegate the development of a regulation to a committee of business executives and representatives of consumer lobbies. Participation by the consumer lobbies and the necessary research by such citizen groups was financed by grants from the CPSC itself. However, unlike Gusman's dialogue groups, the CPSC's offeror process is generally thought to be an almost complete failure, for consumer representatives could not quickly develop a position on the numerous technical issues involved in the manufacture of a safe consumer product.[10]

By 1979, the term "regulatory negotiation" was used by those in the Washington issue-network concerned with the development of new forms of dispute resolution techniques: the Conservation Foundation, the American Bar Association, consultants on mediation of environmental disputes, and so forth. In July 1980, Senator Carl Levin (D-Mich.) held hearings on the concept.[11] The lead testimony was presented by representatives of the NCPP: Larry Moss, Frank Murray, and Harrison Loesch, the leading coal company lobbyist on strip-mining issues. The NCPP's testimony was followed by that of Sam Gusman about the Conservation Foundation's dialogue groups. The idea of regulatory negotiation was praised by representatives of President Carter's Regulatory Council and the EPA. Testimony by the U.S. Chamber of Commerce favored the concept, although that organization had not taken an official position on the matter. Regulatory negotiation bills were subsequently introduced into the Ninety-sixth, Ninety-seventh, and Ninety-eighth Congresses by Senator Levin and Representative Pease. However, no hearings were held on any of these bills. (A bill eventually became law in the 101st Congress in 1990.)

In 1981, a small federal agency, the Administrative Conference of the United States (ACUS) became interested in furthering regulatory negotiation. A little-known body, the ACUS might be described as a group that is designed to act the role in administrative law reforms that the Advisory Commission on Intergovernmental Relations acts in intergovernmental relations. The ACUS is an assembly of ninety-one governmental and nongovernmental attorneys that meets once a year to consider recommended changes in administrative procedures. The ACUS has only a small staff and does much research by contract.[12]

In June 1982, the ACUS endorsed a regulatory negotiation proposal, largely drawn by Philip Harter, viewed as the leading expert in the field.[13] The ACUS's action gave the idea legitimacy in the field of administrative law. Harter's 118-page law journal article, "Negotiating Regulations: A Cure for Malaise," has become the basic text of regulatory negotiation.[14]

Let us briefly consider a number of aspects of regulatory negotiation proposals. Of course no two proposals are identical; proponents of the concept disagree on a number of issues, such as the exact role of representatives of government, whether public funding should support representatives of public-interest lobbies, and so forth.

1. A mixed group of interest-group leaders might assemble and then approach a regulatory agency about establishing a negotiating process. Or an agency itself might organize the group.

2. Such a group would then approach a neutral convenor, usually designated to be the ACUS, to organize the process. The convenor would ascertain whether additional groups should be invited and a professional mediator be hired. The convenor would staff and fund the negotiators from public appropriations. Some proposals would permit public funding for public-interest group representatives, so that they could hire replacement personnel for their understaffed organizations.

3. The negotiators would meet with senior agency personnel to develop a consensus set of regulations about an issue. The number of negotiators would not be large—fifteen is often the recommended number. Consensus or near-consensus would establish agreement. Agency personnel would not necessarily vote at this stage, however.

4. Negotiation cannot succeed on all issues, for certain intensely contested issues (e.g., many related to nuclear power) cannot be decided by compromise. On the other hand, some issues may be subject to compromise, because all sides prefer to avoid the alternative of long years of expensive litigation.

5. Although most negotiation can take place in open meetings, some bargaining will need to take place in closed session. Accordingly, open meetings provisions of the Federal Advisory Commission Act may have to be amended to provide for closed bargaining sessions.

6. Government agencies must be mandated by statute to pay serious attention to the report of a regulatory negotiation council. For instance, affected regulatory agencies may be required to reply publicly to such a report within sixty days. The report of the council also might be sent to relevant congressional committees.

7. In considering the report of a regulatory council, an agency might issue a notice of proposed rule making. Comments and proposed amend-

ments to the regulatory-negotiation report should be reviewed in a conference between the negotiators and the agency. The agency might then publish the final rules.

Not everyone supports regulatory negotiation. One argument against the concept is that it is too complex—regulatory-negotiation procedures might take as long as ordinary rule making. Furthermore, there is no guarantee that one of the negotiators might not break the agreement and resort to litigating the rule, in which case nothing has been gained. Others pointed to the failure of the offeror process in the CPSC.

Another objection suggests that one or more procedures in a regulatory-negotiation proposal are unnecessary, resource consuming, or harmful to the goal of agreement. For instance, some opponents believe that participation by the regulatory agency in negotiation might distort the process. Others see no real need for a neutral convenor; they believe it just complicates the process. Finally, conservatives usually object to public funds being appropriated for consumer representatives.

EARLY EXPERIMENTS IN REGULATORY NEGOTIATION

The National Coal Policy Project might itself be considered an early experiment in regulatory negotiation. The NCPP's process demonstrated the difficulties of implementing a widely varied policy platform, and though it did not get a lot of publicity, the NCPP itself conducted successful regulatory negotiations, if we stretch the term to include the two-sided discussion process between itself and the Federal Energy Regulatory Commission about deregulating cogenerators as public utilities.

In the years 1982–1985, there were five experiments in regulatory negotiation involving the federal government. Four of these negotiations initially seemed to be successful; the fifth had the same problem as the NCPP—the negotiators finally reached agreement among themselves, but the groups they represented, as well as the government agency, refused to support the agreement. The successful agreements resulted from the following: negotiation between environmentalists and the Chemical Manufacturers Association without government participation over a type of PCB contamination; negotiation among twenty-two organizations and the EPA over nonconformance penalties for truck-engine emissions; negotiations among twenty organizations and the EPA concerning emergency waivers to federal pesticide control regulations; and negotiations among the Federal Aviation Administration (FAA), unions, airline companies, and aviation users concerning the amount of rest time required for airline pilots. The negotiation that failed to gain support from the representatives' groups concerned the Occupational Health and Safety Administration (OSHA) attempt under court order to develop standards regarding the exposure of workers to benzene in manufacturing processes.

It is beyond the scope of this work to conduct a thorough analysis of these first five processes of regulatory negotiation; Henry Perritt has analyzed four of these cases.[15] But a few basic observations are in order. The major point is that regulatory negotiation is sometimes successful, though so far we have no idea of all circumstances necessary for success.[16] Nor can we be completely sure that the agreements reached through regulatory negotiation might not break down years after the resultant regulations have been issued and that protracted lawsuits might not then develop.

An early negotiation effort over controlling PCB pollution did not include government officials, and thus was not, strictly speaking, regulatory negotiation. But because its success encouraged later experiments with regulatory negotiation, I include it here. In 1982 representatives of the Chemical Manufacturers Association, the Environmental Defense Fund (EDF), and the Natural Resources Defense Council (NRDC) (which had refused to participate in the NCPP) met to develop regulations designating as impurities PCBs in concentrations below fifty parts per million in chemical manufacturing processes. The negotiations were conducted for about a year; the negotiating parties proposed regulations in April 1983; these were adopted without major change by the EPA in December 1983.[17]

These negotiations over the inadvertent creation of PCBs in manufacturing processes were successful in part because the chemical industry had a great incentive to avoid lawsuits for infringing health and safety with PCB pollution, at the top of the public's list of environmental concerns as a possible carcinogen. The two environmentalist groups, the EDF and the NRDC, were almost certain to sue chemical companies on this issue. The chemical industry needed some rules governing the inadvertent production of PCBs to demonstrate its lack of negligence because specific safety rules were being followed in manufacturing processes.

Environmentalists were motivated to reach agreement, on the other hand, to save litigation expense in a situation in which industry appeared to be exceptionally eager to reach a timely agreement. Richard Ayres of the NRDC told me in 1979 that he preferred to settle out of court if possible because of the expense of litigation. In the PCB case, Ayres did not object to NRDC's participation, apparently because the Chemical Manufacturers Association communicated a desire by top executives to commit their companies to a set of regulations. Moreover, timely agreement was attractive to environmentalists as a means of reducing the emission of PCBs into the environment a few years earlier than normal adversarial processes might provide.

The PCB negotiations had some similarity to the NCPP in that the groups conducted negotiations without government participation. The PCB interest groups reached agreement among themselves and (unlike the NCPP) persuaded a government agency to enact their proposals. The PCB negotiations were unlike the NCPP in that proposals were limited to one quite limited issue. Two environmentalist groups and one trade association participated, which made the PCB negotiations much simpler than those of the NCPP. In addition, the industrial creation of PCB contaminants was an issue ripe for settlement. The environmentalist participants were those who would sue some of the members of the trade association if no agreement were reached. The PCB negotiations had the characteristics of out-of-court settlement; both parties to potential litigation had strong financial incentives to avoid such litigation.

Another reason the EPA cooperated with the PCB negotiators is that the agency's top management was friendly to the idea of regulatory negotiation. The EPA director under President Carter, Douglas Costle, was a friend of Larry Moss and expressed an interest in *Where We Agree*. This link was important because the EPA was the relevant enforcement agency for the NCPP's air pollution proposals. Costle's verbal interest was not, however, translated into directions to subordinates, according to people then working in the EPA's policy planning division. An Assistant Administrator under Reagan, Joseph Cannon, became interested in the idea of regulatory negotiation and gained the backing of Reagan's first EPA director, the now infamous Anne Gorsuch Burford, for the idea. Before his retirement in 1983, Cannon made it the policy of the EPA to search for a problem to which the regulatory-negotiation process could be applied as a solution.[18] After Burford's resignation in 1983, her successor, William Ruckelshaus, expressed support for experimentation with alternatives to the normal adversarial process and even gave the keynote address to a conference of dispute mediators in October 1984.[19] At about this time, EPA officials located two more problems for the regulatory negotiation "solution."

In April 1984, the EPA initiated regulatory negotiations over the question of nonconformance penalties for the manufacture of heavy-duty truck engines that did not meet the emissions standards set by air-pollution legislation. The rationale for these regulations was that it might take several years for some manufacturers to redesign their engines to meet the new legal specifications, and the EPA intended to pressure such manufacturers to obey the law without putting them out of business. Accordingly, the EPA decided to fine such "nonconforming" manufacturers and determined that the nature of these penalties should be set by a regulatory negotiation group. The twenty-three representatives composing this group were affiliated with both small and large truck manufacturers from the United States, Japan, and Europe, with five automotive manufacturing and trucking associations, and included representatives from state pollution control agencies and the Natural Resources Defense Council. Unlike the earlier case of the PCB negotiations, the EPA itself participated in the truck-engine nonconformance negotiations. After intensive work, the group reached consensus on the basic issue by October, and the EPA issued preliminary regulations in March 1985.[20] These preliminary regulations drew thirteen comments, all from participants in the negotiating group, and all were positive.

The second EPA-sponsored regulatory negotiating group in 1984 concerned the issue of granting emergency exemptions to state and local governments to use pesticides that had been controlled by the federal pesticide control act. A multi-sided negotiation representing twenty-one organizations formed, with participation by the EPA, the Department of Agriculture, state agriculture and health departments, environmentalist lobbies, trade associations, and pesticide users. This group was able to agree on the exact wording of regulations between August 1984 and January 1985, and the EPA published a notice of the proposed rule in April 1985.[21] The nineteen comments received by the EPA concerning the proposed rule generally "raised relatively minor points of interpretation."[22]

A major success for the regulatory negotiation process was an agreement among the FAA, organizations of pilots and navigators, airline companies, and aviation users concerning flight- and duty-time rules and rest requirements for aircraft crews.[23] The negotiations process involved nineteen separate interests; it began in May 1983 and ended with the new regulations issued by the FAA in July 1985. The situation was highly complex. Previous FAA regulations about flight and duty time had been rendered obsolete by technological change in the airline industry, but no new regulations took their place, only years of argument among the various concerned parties. Accordingly, most of the FAA's flight- and duty-time regulations had little application to the contemporary airline business. The complexity of decision making is indicated by the subject matter of agreement: how many hours of rest should pilots get per day? per week or month or year? Should this determination vary with different types of flight duty, and if so, how should it vary? Types of flight duty include piloting various types of aircraft on various types of runs: commuter, short hops, transcontinental, and so forth. One reason for seeking agreement was that all parties were frustrated by the many years of endless bickering and looked to the new idea of a multisided bargaining process as a fresh, untried means of devising regulations. Even so, the FAA Administrator had to unilaterally issue some of the regulations.[24]

One of the processes failed—OSHA's attempt to solicit agreement from chemical manufacturers and labor unions over the use of benzene in manufacturing processes. These negotiations proceeded for about eighteen months and yielded in September 1984, a protocol among the negotiating representatives of unions and industrial associations, although OSHA itself did not participate directly. But the chemical manufacturers, the unions, and OSHA all rejected the agreement drawn up by the negotiators.[25]

The benzene negotiations confirmed some of the lessons of the NCPP's experiment. Representatives of the chemical manufacturers and unions were also able to reach substantial agreement in the benzene negotiations, but since OSHA did not take part in the negotiations, the negotiators were not able to translate their agreements into public policy. One problem was that the leadership of OSHA changed during the negotiations, and the new agency chief was not prepared to make decisions about the results of the benzene negotiations.[26]

If we count the NCPP's negotiations in the areas of cogeneration and mine bonding as two instances of successful negotiations, then all five regulatory negotiation experiments including government personnel succeeded. In the two cases in which government officials were excluded, on the other hand, one succeeded (PCB) and one failed (benzene). One difference between the two cases was the attitudes of agency leadership. Perritt reports also that benzene negotiators all came to believe that they could achieve their goals more fully outside of the negotiation process,[27] while the PCB negotiators had strong incentives to avoid litigation among themselves. The initial evidence, then, strongly indicates that including government officials from implementing agencies greatly enhances the possibility of success in regulatory negotiation. Voluntarism was not successful in the NCPP, nor is it likely to be successful in regulatory negotiation.

After its successful relationship with three regulatory negotiation groups, the EPA itself had some general sense of when the new procedures would work: "We learned that selecting an appropriate item [for negotiation] is our most critical step; that unless the time is 'ripe,' the parties and issues identifiable and manageable; and the parties willing to go to the table in good faith to reach consensus; it is not wise to proceed."[28] Most of these adjectives are somewhat vague, but if the indicators could be stated more specifically, a useful test for undertaking regulatory negotiation could be developed.

The EPA continued to experiment with regulatory negotiation. By late 1986, according to Daniel Fiorino, a political scientist and policy analyst of the EPA, five of the EPA's experiments were judged by the agency to be successful (including the three just described), while two were evaluated to be failures. In an article in the *Public Administration Review*, Fiorino stated his conclusions about the seven EPA experiments:

> Within its limits, and with procedural safeguards, regulatory negotiation offers a valuable complement to the conventional process. EPA's experience suggests that negotiation can yield rules that are superior policy products, are based on better information, and are more likely to reflect the informed preferences of affected groups. In addition, negotiation offers benefits as a deliberative process, in an arena where open, purposeful, and reasoned deliberation is rare. Further experience may confirm that regulatory negotiation used se-

> lectively can promote democratic processes and ideals more than conventional rulemaking can on its own.[29]

After a three-year hiatus, the EPA launched four more regulatory negotiation experiments in 1989–1990.[30] In addition to the EPA, OSHA, and the FAA, in the late 1980s the Department of the Interior, the Federal Trade Commission, the Department of Education, the Department of Agriculture, and the Nuclear Regulatory Commission all experimented with regulatory negotiation.[31] At that time, ten states initiated negotiated rule-making processes, and other states seemed about to launch such experiments.[32]

Interest in regulatory negotiation is growing, for it has a great deal of appeal as a rather obvious, but still fresh procedure to develop regulations in a timely manner recognized to be legitimate by those groups affected by the new rules. Obviously, regulatory negotiation works in some situations and fails in others. During the next few years, a sorting process will occur as attempts at regulatory negotiation increase. Experts will gradually learn to identify when the chances are good for successful negotiation.

THE NCPP AND REGULATORY NEGOTIATION

Regulatory negotiation would have been initiated even if the NCPP had never occurred. Still, the NCPP inspired interest in the concept of regulatory negotiation after 1978. Concrete evidence existed in the two volumes of *Where We Agree* that negotiations among hitherto conflicting groups could produce productive agreement. The NCPP's leadership came to the view that one of the major effects of the project was to encourage regulatory negotiation. The *Final Report* states: "Potentially the most significant use of the NCPP's process as a model has been in bills introduced in Congress to create regulatory negotiation commissions."[33] The chief proponent of regulatory negotiation in the House had been Representative Pease, the chief congressional supporter for the NCPP, who cited the successes of the project in his speech introducing a bill to promote the new negotiation process.[34] Senator Carl Levin (D-Mich.) was the chief proponent of regulatory negotiation in the Senate. In introducing a bill similar to Pease's bill, Levin stated: "The most well-known regulatory negotiation success story is probably that of the National Coal Policy Project (NCPP)."[35] The basic text of regulatory negotiation, the legal monograph published by Philip J. Harter in 1982, refers to the NCPP as one of six "analogues" to the new concept, and describes the project as follows: "Perhaps the best known example in which parties that are usually adversaries negotiated numerous agreements concerning public policy was the National Coal Policy Project (NCPP)."[36]

The efforts of Senator Levin and Representative Pease were instrumental in eventually passing the Negotiated Rulemaking Act of 1990. They were strongly committed to the development of regulatory negotiation and therefore introduced bills into each congress during the 1980s and persuaded members holding key committee positions to back the idea. Meanwhile, the idea of regulatory negotiation gained momentum among those concerned about regulatory reform. As I noted, the Administrative Conference of the United States endorsed the idea in 1982. By the beginning of the 100th Congress in 1987, it could be argued as an empirical fact that regulatory negotiation sometimes worked. In 1988, Congress inserted a negotiated rule-making requirement into policy-making requirements for federal aid for the education of disadvantaged children. Also in 1988 Congress stipulated that the Nuclear Regulatory Commission use a form of regulatory negotiation in regulating the medical use of radioactive isotopes.[37] And at the end of the 100th Congress, the Senate passed a general negotiated rule-making bill, but it was too late for the House to act before adjournment.[38]

In October 1990, however, both houses of Congress passed the bill which became the Negotiated Rulemaking Act of 1990. This law did not require regulatory negotiation, but served as a congressional endorsement of the process, outlined suggested procedures, exempted proceedings from conflicting requirements of the federal Advisory Committee Act, authorized small expenditures for the purpose of supporting participants lacking financial resources, and noted that negotiated rule-making procedures were not given special status in the federal courts.[39] No opposition to the act was expressed in the *Congressional Record*. The key House vote was 411–0, with 1 "present."[40] Other votes in the Senate and the House were voice votes.[41] Congressional consideration of regulatory negotiation was more a process of persuasion than a political process reflecting conflicting interests. Levin and Pease were able to persuade a few key actors to support the idea, such as Senator John Glenn, chair of the Senate Governmental Affairs Committee, Representative Jack Brooks, chair of the House Judiciary Committee, Representative Barney Frank, chair of the relevant Judiciary subcommittee, and Representative Craig James, the minority leader of the Administrative Law and Governmental Relations Subcommittee. In the *Congressional Record*, the aggressively humorous Congressman Frank addressed Congressman James, "If the gentleman will yield, he, I, and the gentleman from Ohio [i.e., Pease] are the only people in the House who have read the bill, so if we agree, I think that pretty much locks it up." Congressman James objects, however: "I know that my side of the aisle has read it with great interest and concern."[42]

RECENT EXPERIMENTATION

During 1991 the Bush Administration's Environmental Protection Agency actively experimented with new ways of negotiating with interest groups to devise regulations to implement the Clean Air Act Amendments of 1990. This law was the centerpiece of President Bush's environmental policy. After about two years of intensive negotiations among the administration, Congress, and interest groups, the final bill was adopted by the overwhelming margins of 89–10 in the Senate and 401–25 in the House and was signed by President Bush on November 15, 1990.[43] Among the goals of the bill were tighter regulations on emissions thought to produce acid rain and increased control of emissions producing urban smog and toxic chemicals in the air.

To address the process of consulting with interest groups and local governments about implementing the 1990 Clean Air Act Amendments, the EPA initiated unusually careful and thoughtful planning procedures in December 1990. An internal memorandum recognized that "if at the close of a rulemaking process, significant issues of concern to major interest groups remain . . . it is more likely that the rule will not be implemented on time."[44] The memo discusses a number of different approaches to consulting with interest groups, including regulatory negotiation, but also including formal and multi-interest policy dialogues (meeting under the rules of the Administrative Procedures Act), informal multi-interest policy dialogues or roundtables, and multi-interest workshop meetings of one to three days, as well as meetings with representatives of a single interest sector and traditional public meetings.[45]

The 1990 Clean Air Act set a deadline of November 1991 for the EPA to produce a new set of standards for refining a reformulated gasoline that would reduce ozone pollution as well as carbon monoxide and toxic agents in the air of urban areas. The implementation of such standards would be delayed long after the November 1991 date unless the EPA launched a major effort to gain support from the major interests involved. Accordingly, the EPA decided to use the formal regulatory negotiation strategy to devise such reformulated gasoline standards, and a regulatory negotiation process was established in March 1991 with the process's major theoretician, Philip Harter, as the "facilitator," or leader. Thirty-one entities were represented in the regulatory negotiations: oil companies included Chevron, Exxon, Sun, Amoco, Mobil, and Phillips; environmental groups included the Sierra Club, the Natural Resources Defense Council, and Citizen Action; Ford and General Motors participated; the EPA itself participated, with the U.S. Department of Energy and representatives of the states of California, New Jersey, Colorado, and Rhode Island; representatives of the associations of state air control administrators as well as local air control administrators participated; and petroleum marketers, gasoline stations, independent refiners, and fuel developers were also represented. Because the process of reformulating gasoline standards was thought to involve the expenditure by industry of from three to five billion dollars, as well as an increase in gasoline costs of five cents a gallon in certain urban areas, this particular regulatory negotiation was probably the most significant to be conducted up to that time.[46]

The negotiators announced an agreement on August 16, 1991. According to press accounts, negotiators and the EPA were pleased by the content of the agreement, but the promulgation of rules for gasoline reformulation has been since delayed by the politics of regulation in the probusiness Bush administration.[47] According to the EPA press release on the regulatory negotiation:

> The Clean Air Act requires reformulated gasoline to be used in the nine areas [Baltimore, Chicago, Hartford, Houston, Los Angeles, Milwaukee, New York, Philadelphia, San Diego] with the worst ozone levels. The Act also allows other areas exceeding the ozone standard the option of implementing a reformulated fuels program. The new gasoline will have lower volatility levels, which reduce hy-

> drocarbon emissions (an ozone precursor), and limits on sulfur, olefins, aromatics, benzene and other toxics. Provisions of the Act also exist which prohibit the transfer of benzene and other high polluting compounds removed from reformulated gasoline into conventional fuel sold in the remainder of the country.[48]

The EPA also reported that: "The committee also resolved issues concerning EPA certification of these new gasoline blends and the methods of averaging for volatile organic compounds, benzene, and toxics. Further, the committee developed a survey scheme to assure compliance with the regulations in each area to ensure that the performance standards are fully achieved in practice."[49]

The goal of the standards was to reduce ozone-causing compounds by 15 percent and carbon monoxide emissions by 20 percent in the major urban areas.[50] The participants in the negotiations process signed an agreement, barring each from challenging the agreement through litigation or lobbying.[51]

At about the same time that the gasoline reformulation negotiations were conducted, the EPA encouraged another important series of negotiations between electric utilities and environmentalists over haze-creating emissions from the enormous Navajo Generating Station, located fifteen miles north of the Grand Canyon in Arizona. If effectively implemented, the agreement from the Navajo Generating Station (NGS) negotiations will be a conciliatory landmark in a long-standing, intense conflict involving the environment. During the 1960s, the federal Bureau of Reclamation proposed building two dams in the Grand Canyon to generate hydroelectric power, but this proposal was blocked by the Sierra Club and its allies in a classic environmental battle. After the hydroelectric project was defeated, a group of regional utilities and the federal Bureau of Reclamation formed a consortium to build a coal-burning generating station in the area, the Navajo Generating Station. Once constructed, the 2,250-megawatt NGS was one of the biggest and probably the most notorious coal-burning power plant in the country. In 1991, when operating at full capacity, the NGS burned 24,000 tons of coal per day, and sulfur dioxide in the emissions was marring the beauty of the Grand Canyon by creating haze.[52] "In recent years, thick haze had turned the Grand Canyon from a breath-taking natural wonder to a sad tableau of spreading industrial pollution and Government inaction. On many days, particularly in the winter, the haze is so thick that visitors standing on one rim can barely see the other side."[53]

Not surprisingly, in 1982 the Environmental Defense Fund and the National Parks and Conservation Association filed a joint lawsuit to enforce the Clean Air Act. Litigation dragged on for years without settlement, partially because one agency of the Department of the Interior, the Bureau of Reclamation, owned 24 percent of the NGS and opposed the imposition of strict emissions controls. Meanwhile, another unit of the Department of the Interior, the National Park Service, vigorously supported the cause of limiting sulfur-dioxide emissions.

In 1990, Bush's EPA administrator, William K. Reilly, was prepared to order a 90 percent reduction in sulfur-dioxide emissions at the NGS. But the Office of Management and Budget objected, saying this cost the utilities too much, and thus in February 1991, Reilly proposed a 70 percent reduction. Reilly's tough proposal gave an incentive to the utilities to negotiate. Reilly proposed a 70 percent reduction each month, much more expensive than a 70 percent reduction averaged over the entire year, because the monthly quota necessitated the purchase of additional expensive scrubbing equipment to provide immediate back-up emissions control in case of a shutdown of an operating scrubber.

At the behest of the EPA, the two sides began negotiations in April 1991. The negotiating body was not established as a formal regulatory-negotiation process, but had only the status of an informal advisory body to the EPA. The NGS's consortium of owners was represented; the Salt River Project Agricultural Improvement and Power District, operators and 22 percent owners; the Bureau of Reclamation, 24 percent owners; the Los Angeles Department of Water and Power, 21 percent owners; the Arizona Public Service Co., 14 percent owners; the Nevada Power Co., 11 percent owners; and the Tucson Electric Power Co., 8 percent owners. Environmental negotiators represented the Environmental Defense Fund, the Grand Canyon Trust, the Wilderness Society, the National Wildlife Federation, and the Sierra Club. While the EPA was not a formal participant, it facilitated the negotiations by providing technical specialists and data.[54] EPA officials, who asked to be unidentified, told the *New York Times* that the White House first opposed EPA's facilitation of the negotiations, but "later relented after the Bush Administration recognized the symbolic political importance of a deal to scrub the skies in the Grand Canyon."[55] Obviously TV spots of the ruination of the Grand Canyon's beauty might have been an effective weapon against the Bush reelection campaign in 1992; Lee Atwater had used this very weapon in 1988 by running television spots showing the despoliation of Boston Harbor and blaming it on Governor Dukakis.

On August 8, 1991, the EPA announced that the negotiators had reached an agreement based on trading a 90 percent reduction in sulfur-dioxide emissions for the annual averaging method of determining emissions. This method was expected to cost $430 million for the purchase of scrubbers, rather than the $510 million needed for the monthly averaging system. Construction of the scrubbers would take several years, and the new system of emissions control was not required to operate until 1997–1999, when it would be gradually phased in. In addition, maintenance shutdowns were to be concentrated in the winter, when climatic conditions produced the worst haze.[56]

Unlike other cooperative negotiations conducted by the EPA in 1991, the Grand Canyon agreement was speedily implemented because the White House believed it would benefit the President's political fortunes. The EPA pronounced it "largely adopted" in rules promulgated on October 3, 1991.[57] President Bush had already praised the agreement on September 18 in a telegenic speech delivered in front of the Grand Canyon.[58]

A third significant cooperative negotiation experiment was conducted by the EPA in 1991 to assist in gaining widespread support for a system of implementing the acid-rain provisions of the Clean Air Act of 1990. The proceedings of this Acid Rain Advisory Committee (ARAC) were advisory and thus did not constitute a regulatory negotiation process. Nevertheless, according to an EPA information sheet for the public: "ARAC convened six public meetings with hundreds of participants. The input received through this process was critical to the development of the core rules of the Acid Rain Program."[59] Apparently the deliberations of this group substantially influenced the EPA's initial design of a system of rules for reducing the sulfur-dioxide and nitrous-oxide emissions considered to cause acid rain.

Expecting the passage of a Clean Air Act, the EPA issued a public invitation in August 1990 to participate in the ARAC. Interest in participation was widespread; there were 150 nominations for an eventual 44 positions on the committee. As the convener of the committee, the EPA selected representatives from eight constituencies: the electric utility industry, public utility commissioners, state air pollution control officials, environmentalist lobbies, consumer lobbies, manufacturers of pollution control equipment, the coal industry, the natural gas industry, and academic experts.[60]

Unlike the NCPP, EPA officials participated in the ARAC deliberations and the ARAC also reported to this government agency. But the ARAC resembled the NCPP in one interesting aspect; its members were able to agree on the application of market concepts in making recommendations to the EPA regulation writers. Participants in the ARAC generally supported the concept of granting quantitative pollution allowances to coal-burning utilities and other major polluting sources. In the recommended system, pollution allowances could be traded in an EPA-supervised market, as those companies not using up their allowances could sell the surplus to companies exceeding their allowances. The pollution allowances could also be banked against future needs. As the NCPP's discussions revealed, the attractions of a market system of pollution control derive from its cost effectiveness and flexibility. Each company could decide how to regulate emissions at each of its separate plants and could itself determine for each plant whether it might be cheaper to buy scrubbers or other equipment, to purchase pollution allowances from other companies, or to select various combinations of purchasing equipment and allowances.[61]

In conclusion, the year 1991 saw various types of cooperative pluralist negotiations and deliberative proceedings assume major importance in the policy-making process at the EPA. The policies considered by the negotiators were significant, involving allocations of billions of dollars yearly. Currently, administrators of the EPA regard these cooperative negotiations as successful,[62] and the EPA's experiments will probably continue. As supporters of the NCPP hoped in the early 1980s, the record of the NCPP has contributed to a legacy of experimentation with cooperative negotiation processes, as exemplified in recent intergroup deliberations on the enforcement of the Clean Air Act.

9

Cooperative Pluralism

The NCPP deserves to be remembered as a case study of interest group behavior. As long as interest groups are important in America and other democracies, a question will naturally recur to politicians, citizens, and civil servants: what would happen if groups negotiated public policies among themselves, without relying on the state? The NCPP was an elaborate search for an answer to this ever-recurring question, and it leads to an important conclusion. Even if normally adversarial parties find common interests, the resultant agreements are still unlikely to lead to public policies unless the government plays an active role in shaping or implementing agreements.

Although a group of representatives of adversarial groups may reach agreement on some issue or issues, such an agreement will very likely be blocked by a minority veto. It takes only a minority of decision makers within one or a very few groups to veto the agreement of the representatives. One or two of the participating groups are likely to have a minority of decision makers who oppose the joint platform as too compromising of the group's integrity or interests. The platform of common interests, discovered by joint discussion among representatives of the various groups, is not likely to be seen as sufficiently important for group leaders to override minorities opposed to the platform. Such minorities have the potential, for instance, to withdraw from a voluntary group or to oppose group leadership on other issues perceived by the leaders to be more vital.

Furthermore, if one or two of the major groups or companies on a negotiating side denounce the agreement, other organizations on that negotiating side are likely to withdraw from the agreement to avoid offending the opposed organization. Group leaders are unlikely to view the joint compromise platform as sufficiently salient to risk alienating allies whose support would be sought on quite a number of other lobbying issues. In addition, even if one negotiating side maintains unified support for the joint agreement, the compromise position fails if it is vetoed by one or two groups on the other negotiating side. This gives the vetoing organization more leverage to block the agreement, because the vetoer needs to influence only those organizations on its own negotiating side.

The example of regulatory negotiation, however, suggests some potential to develop gradual learning and cooperation among adversarial interests. Such a development, however, apparently requires the presence of the state to initiate the cooperative process and to provide authority for the implementation of any agreements reached by adversarial groups. In other words, the presence of the state is needed to facilitate the learning and implementation of common interests among adversarial group organizations.

The history of the NCPP demonstrates the importance of examining the three-sided interactions of producer groups, countervailing power groups, and agencies of the state acting in the policy process. In this case, the attempt to take the state out of the policy process failed to produce significant results. The idea of some of the early pluralists—that groups were the major influence upon policy, and that government had little independent effect on policy—was proven false in this case. In addition, because of the significant influence of environmentalist lobbies, the "iron triangle," or subgovernment coalition of cooperating producer lobbies, government agencies, and legislators, does not completely control policy in areas related to coal. Consequently, the experience of the NCPP confirms the importance of examining the interactions of independently acting producer groups, countervailing power groups, and government agencies to understand public policy.

One must be cautious, however, not to overgeneralize from the case study of the NCPP. In the American political system, a number of related situations involve bargaining and negotiations among clusters of interest groups, which may be negotiating either with or without direct participation by government officials. Some of these negotiating situations resemble the NCPP at first glance, but on further analysis, they are revealed to have a different structure. Their outcomes may have more direct impact on policy than the NCPP had. For instance, in his study of the passage of amendments to the basic law regulating pesticide usage, Christopher Bosso describes a case of negotiation that partially resembles the NCPP. A representative of the National Agricultural Chemicals Association (composed of ninety-two businesses) met in 1985–1986 with a representative of the Campaign for Pesticides Reform, a coalition of forty-one public interest groups and labor unions, to draft a consensus bill. Government was not represented in these generally one-on-one negotiations. This process was, then, voluntaristic; but unlike the NCPP, it was not participatory, as it was essentially conducted by two persons. More importantly, these pesticide negotiations were focused on a more limited range of regulations than was the NCPP's wider-ranging dialogue.[1]

In this case, the two sides agreed upon a bill, and it received the support of both lobbying coalitions. But some smaller chemical companies and most agricultural groups were left out of the bargaining process between the two coalitions, and for two years these groups succeeded in blocking the passage of the initial joint proposal of the lobbying coalition, which, when eventually passed, had to be revised and limited. Nevertheless, the congressional passage of the limited set of amendments can be viewed as an indirect effect of the negotiating process between the two opposed coalitions. Bosso's case study also illustrates how a minority of interest groups actively propounding a specific issue can greatly affect the reshaping of a package proposal initiated by a cooperative coalition of a larger number of normally adversarial groups.[2]

DISCOVERING COMMON INTERESTS

A second conclusion from the NCPP experiment, less clear than the one above, is that the breadth of common interests among normally adversarial groups may be much greater than is normally suspected, especially if political institutions are constructed to develop such common interests. This is not to say that common interests will dominate the conflicting interests among producer groups and countervailing power groups—only that such interests are an important dimension of group relationships.

Political scientists studying American politics have neglected the study of the common interests among contending interest groups. This situation is reminiscent of international relations theory, in which the realist school, led by Hans Morgenthau in the 1950s, argued that international politics should emphasize the study of the search for power to serve national interests.[3] Many political scientists were trained to accept this outlook, which criticized an "idealism" emphasizing the possibilities of international peace treaties and strong international organizations without considering the conflicts of interests among nations.

A later view modified the realism of Morgenthau. Thomas Schelling and other writers influenced by game theory argued that international conflict is "mixed motive," and that even adversaries usually have important interests in common.[4] Game theorists argued that conflict is not always "zero-sum," in which interests are completely opposed, but is also "non-zero sum," in which both adversaries gain from certain outcomes of an interaction. Schelling extended this idea to analyze the joint interests of the United States and the Soviet Union to prevent Cold War crises from escalating into nuclear war. By now we realize that the superpower adversaries frequently shared common interests in such matters as preventing the spread of nuclear weapons to third parties, communicating about military accidents such as ICBM's veering off course, and so forth.

The experiment of the NCPP demonstrates that while we may initially accept Morgenthau's view of power, the study of interest groups also profits from the game theorists' understanding of mixed-motive, non-zero-sum rivalry among adversaries. Such an assumption is not an idealistic assertion that cooperation is more basic than conflict; it is simply a complex realism that finds that common interests are still present in adversarial relationships.[5] This viewpoint characterizes cooperative pluralism. Recognizing the dimension of common interest in the interaction of producer groups, countervailing power groups, and autonomous agencies promotes the design of such institutions as the NCPP to further the discovery of common interests.

The story of the NCPP offers us five instructive generalizations about the discovery of common interests within a context of normally opposed interests:

1. The NCPP demonstrated to some specialists in dispute resolution, Washington lobbyists, and politicians that the possibilities for agreement among "ambassadors" of adversarial groups is greater than they had previously thought.

2. The NCPP encouraged interest in regulatory negotiation to discover common interests and to affect public policy.

3. The NCPP indicated that a negotiation conference among adversaries within an issue network could improve communication among those adversaries, conceivably leading to a new understanding among opponents of their shared interests, which might have the long-run effect of reducing some types of political conflict.

4. The NCPP demonstrated the role of a culture of negotiation in promoting the realization of common interests among group adversaries. In this case, leadership, organizational structure (similar to the House Appropriations Committee), and the various norms of the Rule of Reason, science, and economics promoted agreement.

5. The NCPP's proposal of most general interest is that licensing hearings for economic development projects be consolidated, and environmentalist groups receive public funding to develop research for testimony at consolidated hearings. This idea represents the spirit of the NCPP: both sides regard the activities of the other as legitimate, and both sides are committed to the timely development and communication of technical data pertinent to regulatory issues.

The National Coal Policy Project involved expenditures of 1.4 million in 1977 dollars and 15,000 days of work. It is not likely that the next few years will witness another private effort of this magnitude to mediate conflict among interest groups over issues of national policy. It is conceivable, however, that in some future era in which cooperation among groups is stressed, the federal government might act as a convener for another mediation group similar to the NCPP. If a mediation conference were established by leading policy-makers, it is likely that any agreements would receive more direct and immediate attention from the national government than did *Where We Agree*.

CITIZENSHIP

Another dimension to the idea of cooperative pluralism is its normative aspect as a theory of democratic citizenship. As I mentioned in Chapter 1, John Stuart Mill advocated an idea of citizenship as political participation in which the participants learned from one another, particularly about common interests. Mill also contributed to what is now known as public choice theory by arguing the advantages of proportional representation as a voting scheme to aggregate interests in a just fashion.[6] Mill thus analyzed interests not only in terms of systems of their aggregation among individuals, currently the dominant analytical mode in political science, but also in terms of how citizens came to learn their own interests.

William K. Muir applied Mill's theory to his own empirical study of the California state legislature. Muir argued that present political economic modes of legislative analysis must be supplemented by a model in which legislators are seen as students, interested in learning and teaching one another about public policy and concerned to advocate and to gain the adoption of policies to serve the general public.[7] Muir's view of the legislature, or at least of the relatively effective legislature of California in the 1970s, is similar to cooperative pluralism. The legislators are influenced by countervailing interests, but most of the outstanding legislators also act autonomously from a personal desire to design and to pass better public policy (an idea analogous to agency autonomy).

The NCPP helps us to see that ideas such as Mill's and Muir's can be applied in the arena of interest group behavior, especially to an important type of behavior involving negotiations among economic producers, political lobbies related to social movements, and autonomous agencies. Cooperative pluralism views both representatives of producer groups and countervailing power groups as citizens who not only advocate interests and seek to embody them in public policy, but who also might learn from one another about desirable public policy, become more aware of common interests even among adversaries, and even perhaps gain a sort of wisdom about politics through political participation. These were the goals of Decker, Whiting, Moss, Murray, and others who set up the NCPP, though they were secondary to the more pragmatic aim of devising public policies which would enable America to compete in international trade and to keep down energy prices at home, while preserving environmental values. As the NCPP proceeded, however, participants came to value more and more what they learned through participation, chiefly what they learned from the "opposition." Consequently, after the project ended, most of its leaders believed they had learned much about public policy in coal-related areas and saw the opposition as having common as well as conflicting interests. Indeed, a follow-up survey by the CSIS staff in 1982 indicated that members of the NCPP's plenum felt that their communication with the opposing side had improved because of participation in the project. In response to the question, "Has there been any change in your relationship with those from the other side since you participated in the NCPP?" Murray and Curran reported:

> Nearly everyone surveyed indicated that their participation in the NCPP changed attitudes on both sides, primarily by better educating them about the true needs and concerns of the other side. Many said that this change in attitude was the most important accomplishment of the NCPP. Examples given of the practical effect of these changed attitudes were the ability to call on others for specific information and advice and improved contacts with the other side in general. The consensus of those answering was that these effects were mutually enjoyed by all participants.[8]

Clearly, the participants themselves, like Muir's California legislators, found the NCPP to be a worthwhile education.

In the normative perspective of cooperative pluralism, it is valuable to maintain institutions that produce wiser citizens (as interest group leaders), even if such institutions seem to have little direct impact on public policy. From this standpoint, the NCPP achieved considerable success despite the fact that so few of its proposals were adopted. Cooperative pluralism as citizenship would be an impractical or unrealistic value if it seemed, however, that establishing such institutions would be extremely costly or would undermine the conduct of effective public policy. But experimentation with such citizenship institutions is not, in fact, particularly expensive.

One might question why judges should be considered the only elite who increase their wisdom by participating in institutional debates and discussions about public policy. Muir, originally a specialist in public law, argued that in the best conditions, legislators could also develop wisdom. Similarly, the NCPP indicates the possibility for interest group leaders to expand their understanding of public policy by participating in cooperative pluralist institutions.

CITIZENSHIP AND ISSUE NETWORKS

Following the introduction of the concept by Heclo in the late 1970s, interest group leaders are now seen as participating in issue networks, communications nets of those concerned with a particular public policy (such as strip-mining regulation).[9] Consequently, concern for developing the citizenship of group leaders reflects a concern for developing new relationships among adversaries in an issue network. In particular, the NCPP emphasized issues in which countervailing power groups, associated with social movements such as environmentalism, opposed producer-oriented groups within issue networks. Examples of movement lobbies also include lobbies for consumers, good government, peace, minorities, and women. More conservative movement lobbies include those for taxpayers, opposing abortion, and advocating prayer.

Producer groups and movement lobbies communicate within issue networks, but such communication is often limited to litigation or lobbying battles. Seldom does communication arise directly between one person and another, unmediated by the institutions of political conflict. Elaborate evidence gathered from a study by Laumann and Knoke of communications networks among lobbies indicates that similar lobbies (e.g., producers of the same good) communicate much more frequently than very different groups, such as economic groups and public interest groups in the same policy areas. Their evidence, gathered by interviews, corroborates impressions provided by the NCPP in which participants indicated that they had no previous personal relationship with someone on the "opposite side."[10]

In the current era, business advocates and movement lobbyists meet most frequently in the courtroom in a system of adversarial relationships. As Wessel and the NCPP's leaders pointed out, litigation proceedings can lead to a suppression of evidence damaging to one's own case, and opponents usually develop legal arguments emphasizing the correctness of one side and the grave mistakes of the other. Neither side has much concern to understand the point of view of the other if a case comes to trial. Nor do litigants normally show much interest in learning from one another about better public policy. Judges alone are expected to develop wisdom to be used in pursuit of common interests. The point is not that the adversarial system of litigation must be changed; the ideal of citizenship of cooperative pluralism implies that adversarial relations be supplemented by institutions of mutual education in pursuit of common interests among adversaries in an issue network.

Environmentalist lobbyists Dunlap and Ayres feared that the NCPP was just a means to persuade environmentalists to give up hard-fought political and legal victories. In general, movement supporters may object that institutions of cooperative pluralism would mainly serve the interests of the "establishment" by co-opting countervailing lobbies. Without outside, independent funding for participants from citizens' groups, such outsiders' lobbies could spend a high percentage of their personnel resources in participation in cooperative experiments. Unfortunately, it is to some degree true that citizens' groups literally can't afford to search for common interests with their opponents.

Yet, some American social movements have been quite extensively organized and are characterized by a division of labor among movement activities and groups. While the Leadership Conference on Civil Rights negotiates compromises in civil rights bills, black protest groups are still available and free to picket.[11] Contemporary American social movements usually have both an anti-institutional protest aspect and a more conventional lobbying and litigation aspect. Lobbyists and lawyers may accept some of the institutional norms of the political system, but most observers would agree that such an acceptance does not undermine all of the goals of the movement (unless it is a truly revolutionary, antisystem movement). It should often be possible for some of the movement lobbyists or negotiators to participate in institutions of cooperative pluralism without compromising basic movement goals. Because cooperative pluralism is defined by the existence of countervailing interest groups, this outlook assumes the continuation of adversarial relations in other aspects of the producer and countervailing power groups.

In interviews, both Louise Dunlap of the Environmental Protection Center and Grant Thompson of the Conservation Foundation suggested that efforts like the NCPP might best be directed at new agenda items. Research is constantly indicating new environmental issues for public policy. Both of these environmentalists pointed out that discussion of issues among adversaries might be more productive before opponents had adopted definite policy positions.[12] This argument is persuasive and has such analogues in international relations as the negotiations internationalizing Antarctica in 1960 or the recent conferences concerning the use of mineral resources lying on international seabeds. In addition, if corporations have not yet developed a position on a particular future issue, they cannot co-opt movement participants to a nonexistent position. Discussion of future issues dovetails with the ideal of cooperative citizenship among groups, as both sides would be engaged in educating the other about perspectives and interests about newly emerging issues.

If cooperative pluralist institutions were to develop within an issue network, educational effects of participation would sometimes spread beyond the circles of lobbyists, negotiators, and chief executives or organizations. This ripple effect apparently did not occur in the case of the NCPP, but repeated participation in other cooperative institutions, such as regulatory negotiation or dialogue groups, would lead participants to inform fellow corporate executives, organizational board members, and rank-and-file members about common interests, new views of "the facts," and joint policy proposals. The resulting, more complex appraisals of one's opponents and public policies would for some citizens increase the level of tolerance for simultaneous conflict and cooperation with adversaries and mark an advance in democratic citizenship in line with the research of Sniderman and others.[13]

As we can see in the concept of regulatory negotiation, the state ordinarily would initiate and facilitate the conduct of institutions of cooperative pluralism. A continuing effort by a government agency to sponsor such interactions among producer and countervailing power groups could raise the level of democratic citizenship within an issue network. Of course, as the NCPP demonstrated, participation by a government agency seems necessary for the implementation of joint policy proposals.

Private foundations might be particularly interested in funding experiments to develop new types of institutions for cooperative pluralism. Surely some could be invented beyond those mentioned in this study: general policy conferences (the NCPP), regulatory negotiation, policy dialogues, environmental mediation, and mediation as a general concept. While limited funding is available for experiments in mediation, one of a variety of activities termed "alternative dispute resolution," the idea of cooperative pluralism places more emphasis on continuing conflicts than do researchers in the field of alternative dispute resolution.[14]

MARKET VERSUS NONMARKET VALUE SYSTEMS

Since the early 1970s, economics has assumed a greater role in the discussion and evaluation of public policies. Most economists emphasize the importance of the market mode of decision making, and thus in many policy areas there has been increased emphasis upon deregulation to restore market criteria, cost-benefit analysis, assessment of impacts of government policies on the economy as a whole, policy evaluation using econometric techniques, market-set prices with direct government transfers to the poor rather than special prices, proposals to incorporate marketlike administrative techniques such as "bubble" proposals to decentralize pollution control to plant managers, and so forth.[15] This increased emphasis upon economic techniques to evaluate public policy is unlikely to change in the foreseeable future.

But as we can see in the history of the NCPP, economic modes of reasoning regularly confront other values for appraising public policy. It may be difficult to define a working market, as is often the case in environmental policy. Pollution yields externalities; who will pay for these? How much should they be charged? How can we set a price for clean air, for instance? Commitment to social justice or to Christian charity may lead one to advocate social welfare payments to the poor. How do such commitments square with economic reasoning? Everyone has heard of such economic and moral dilemmas as spending tax money to keep alive a patient with no hope of recovery, too little welfare leading to miserable living conditions but too much causing "welfare addicts," and so forth. There is a role for economists to advance their line of analysis, and there are roles for others to advocate interpretations of environmental harmony, the just society, racial integration, and other noneconomic values. Finally, there is a place in policy discussions for those who search for common interests among those advocating economics as the basis of public policy and those holding nonmarket-oriented values.

Normally one expects business to offer a promarket interpretation of a policy situation and environmentalists normally to advocate other values as preeminent. Given this natural ideological division, it is desirable for interest group leaders in an issue network to develop some mutually acceptable compromises between market and nonmarket values. This rapprochement is particularly important in the environmental and energy policy areas, which involve highly complex technical issues as well as divisions about political values. It is extremely difficult to forge enough agreement to enact legislation in the air-pollution control and pesticide-regulation areas, for example, due to the great number of interest groups involved and the ambiguity of some of the scientific issues.[16] An ability to communicate about combining market and nonmarket values in public policy enhances prospects for timely legislation that might be interpreted and implemented with stability.

The process of the NCPP did exhibit such a capacity to communicate about market versus nonmarket values and to blend them together in public policy proposals. A sort of mutual instruction in economic and noneconomic values occurred, which led to greater agreement on policy issues. In this case, government officials were not part of the process of learning common interests, but in regulatory negotiation or in other institutions, officials can be made part of the process. As presidential administrations change, new government officials could be acquainted with the agreements previously reached. Indeed, the new officials would be likely to know of the expanded "zone of agreement" from their previous activity within the policy area.

COOPERATION AND ECONOMIC DEVELOPMENT

Since the last months of the Carter administration, American elites—politicians, journalists, scholars—have discussed a possible need to develop new institutions and policies, reflecting common interests of groups, to enhance economic growth. This discussion has been conducted under such rubrics as "special interest power," "the zero-sum society," "industrial policy," and "competitiveness."[17] It has focused on the need to increase cooperation among government, business, labor, academia, environmentalists, and other groups.

Cooperative pluralism does not necessarily imply any of the usual policies advocated under the terms industrial policy or competitiveness. However, it does favor the creation of institutions to facilitate the learning of common interests. Although knowledge of common interests would then be helpful to those who desire greater cooperation for economic growth, such understanding can also be an end in itself.

Pleas for cooperation between business and labor or other groups in the interest of promoting growth have been quite frequent in recent years, as the reader is surely aware. Such pleas refer not only to America's national competition with Japan and other exporters, but also to the economic growth of states or cities, usually in competition with other states or cities. Calls for cooperation normally refer first to business, labor, and government, and sometimes to academia. Environmentalist groups, public interest groups, and property owners opposed to new economic development projects are infrequently mentioned. But this lack of reference is usually due to the vague nature of proposals for cooperation. The advocates of cooperation seldom think through the details of their proposals, including the detail of exactly whose cooperation is needed for growth.

To cite a typical example of the 1980s rhetoric of cooperation, one might have read in the *New York Times* a report of the inaugural address in January 1987 of the newly elected governor of Pennsylvania, liberal Democrat Robert P. Casey. "The first and most important item on his agenda, he told a crowd that heard his inaugural address outside the green-domed Capitol here (Harrisburg), would be the creation of a public-private economic development council. It would be empowered to direct development funds to people and places that needed them most and to industries whose future was most promising." The *Times* notes that Casey's predecessor, Republican Dick Thornburgh, relied on the state's Department of Commerce for economic growth policy. "But Mr. Casey, an old-style liberal Democrat, is prepared to jettison the government-directed effort in favor of a partnership among government, business and labor."[18] The argument here is that specific institutional means need to be designed to find common interests among these three partners, whose goals often conflict. Currently, Governor Casey and the hundreds of other leaders and social commentators advocating such councils are just beginning to realize that specific procedures are needed to find such common interests.

One type of appeal advocates sectoral cooperation among groups. The NCPP was an experiment in such cooperation. Decker, Moss, Abshire, and others advocated participation in the NCPP in terms of the need for common action in the face of American dependence on energy imports. The NCPP consisted, then, of procedures to advance mutual learning of common interests to enhance the economic efficiency of the coal sector while preserving environmental values. The NCPP furthered the development of other institutions—regulatory negotiation, the licensing tradeoff procedure—by which common interests could be learned in the cause of promoting economic growth.

Advocates of cooperation for growth must pay considerable attention to the means of bringing environmentalist groups into institutions for cooperation. Environmentalist groups demonstrated their continuing power during Reagan's first term, in which such appointees as James Watt and

Anne Gorsuch Burford decreased governmental activity to maintain the environment. After the EPA budget was cut, the resulting political uproar checked the Reagan Administration's policy direction. As far as one can predict, the power of environmentalist lobbies is permanent, for it is based on organizing widespread political values, a continuing willingness to contribute to prevent environmental degradation, and a good understanding of lobbying and litigation techniques.[19] There was a great deal of hostility between environmentalists and business groups in 1976 as the NCPP was organized, but this may have lessened outside of the NCPP as the two kinds of lobbyists gained more mutual knowledge. Yet the political combat initiated by Reagan, Watt, and Burford exacerbated a countertrend of hostility and suspicion on the part of environmentalists toward government and business leaders. Special attention is needed, then, to the development of institutions to teach environmentalists, business, and government about their common interests, if one advocates the platform of cooperation for economic growth.

THE FUTURE OF COOPERATIVE PLURALISM

To what extent have American political institutions seized upon cooperative pluralism to solve knotty policy issues? At the federal level, the extent is apparently small. Why so?

Washington interest-group politics is dominated by hired agents. Interest groups employ lobbyists to represent them. Such lobbyists normally live in Washington and are found in law firms, so-called public relations firms, and in particular entrepreneurial enterprises specializing in lobbying. The best available research on the social structure of the lobbying community—conducted by Edward Laumann, Robert Salisbury, David Knoke, and their collaborators—reveals little interaction among business lobbyists and environmentalist or consumer lobbyists.[20] The same research shows that very few lobbyists are concerned with a large range of issues in a general policy area such as energy, health, or agriculture, with labor policy being some exception. Washington lobbyists, then, are splintered into groups of like-minded interests, who interact mostly with those with similar interests, which tend to be rather specific (natural gas policy, for example, but not energy policy).

Interest-group politics in Washington is not the politics of citizens, all of whom reside in the same city and many of whom feel a sense of identity and loyalty with the community as a whole. When urban citizens represent opposed interest groups, they sometimes do respond to appeals to work together for the joint welfare of their political community. Washington lobbying, on the other hand, is a politics of exchange relationships among hired agents. Lobbyists are hired to represent interest groups, and such lobbyists need not be loyal to anything or anyone; they are required only to honor a contract for representation. The lobbyist then attempts to gain support for the goals of the contractor by communicating with legislators, other interest-group representatives, executive branch officials, or the press, using direct persuasion or offering inducements, normally legal ones such as campaign contributions or helpful publicity.

In short, Washington interest group politics is conducted within a splintered social structure, in which like associates with like, consisting of processes of bargaining over relatively narrow issues among the hired agents of interest groups. Such persons, interacting in roles prescribed by contracts, are not likely to be moved by appeals to cooperate to find common interests, for they have no motivation to do so. Accordingly, there has been little movement to establish cooperative pluralist institutions at the federal level.

One might expect that the extraordinary financial pressures upon cities today would enhance the growth of cooperative pluralism, as city government, business leadership, and countervailing power groups such as organized labor, environmentalists, or African-American organizations work together to promote local economic development. This is done, for instance, in cooperating to promote industrial relocation to a city by agreeing to a program of special subsidies.[21]

Similarly, one might expect that fiscal pressures would induce the state government, business leadership, and labor to cooperate to maintain and develop a state's industrial base. And, indeed, a number of leading scholars of state government are conducting research into this question. This research is now unfinished, but though early conclusions indicate some cooperative pluralism in state government economic development policy-making, the trend as of 1990 was not strong.[22]

Just as it makes sense to project an increase in cooperative pluralism at the state and local levels due to fiscal stress, so certain areas of federal policy-making may develop cooperative pluralism due to economic pressures. Unfortunately, it is very plausible that the American economy will in general remain sluggish and have difficulty competing in many international markets. More specifically, continuing concern about the economic effectiveness of federal policy is likely in certain issue areas, such as environmental policy, energy policy, and health policy. In addition, many people will be concerned about the enhancement of environmental or egalitarian values in these areas. Great difficulty in policy-making is likely to ensue as values of efficiency, equality, and environmentalism clash.

Such difficulties are likely to be prominent in the implementation phase of such policy areas as environment, energy, and health. Let us assume that Congress is able, upon occasion, to pass general legislation to address problems in these areas, and that at times the president is willing to sign such legislation. Even so, the implementation of new laws is likely to run into a thicket of well-organized, opposed interest groups in areas such as environment, energy, and health. Groups will seek to delay implementation of laws by continual, protracted litigation, supervised by numerous able lawyers. Groups will continually seek to delay or appeal the application of regulations by appealing to the congressional amending process, to other competing executive agencies that might support some group's point of view, to the Office of Management and Budget, or to the White House staff.

Still another difficulty occurs when state and local governmental units are incorporated into the federal regulatory process, as in the state implementation programs under federal clean air legislation.[23] It seems possible that state and local governments and newly created local health councils will be incorporated into the implementation of a future national health insurance program. In the environment, energy, and health areas, however, Congress will not be able to foresee all important policy issues and issue clear legislative standards. Accordingly, the implementation process is likely to be especially important in these issue areas, and much of this implementation will be delegated to local councils. But policy could be subject to great delay as interest groups, well organized at both the national and the local levels, strive to influence the implementation of policy at the local level.

Of course, more problems with environment, energy, and health policy arise if Congress cannot pass significant legislation due to political deadlock among interest groups, even though a preponderant majority of the informed public wishes a policy change. When this situation occurs, say in modes of implementation of cost controls of federal expenditures to treat kidney-dialysis patients, one outcome might be that local health councils or state governments would sometimes negotiate a policy for themselves through mechanisms of cooperative pluralism. In other words, future environment, energy, and health policy might feature congressional inaction leading to a delegation of decision-making authority to local administrative bodies, some of which might adopt the practices of cooperative pluralism.

At the federal level, the development of cooperative pluralism will take place within rather specific issue networks, such as lobbyists, government experts, and business executives concerned with strip-mining or the burning of coal. The experience of the NCPP has shown that within issue networks normally opposed lobbyists, after cooperating to find common ground, are capable of learning from one another, communicating with one another, and possibly initiating new cooperative exchanges. In this way, the NCPP demonstrates that cooperation among adversaries need not be limited to a single idiosyncratic decision, but can possibly develop into a recurring phenomenon on a number of decisions involving the same issue network.

Environmental lobbyists will be present in many issue networks. In the health area, to give another example, advocates for the disabled, minorities, and the poor are likely to bring social movement affiliations into issue networks. Social movements are a source of countervailing power groups, but radical, uncompromising movement representatives sometimes present obstacles for agreement on new public policies. The NCPP showed, however, that it is possible for adversaries, attempting to cooperate on an issue, to develop a new outlook combining market and nonmarket values, thereby facilitating agreement.

Public policy under cooperative pluralism will be more responsible to the electorate than a policy characterized by the control of administration by single-interest cliques or a policy characterized by deadlock among contending groups. This system of government affords a greater chance of expeditious and effective implementation of the law as passed by the legislature, as contending groups and government agencies seek to find common ground and no single interest sector dominates implementation or vetoes regulations needed to enforce a law.

In considering the prospects for cooperative pluralism, the crucial lesson of the National Coal Policy Project must not be forgotten: cooperation will not work unless government agencies take part in the negotiations.

Appendix A: Participants in the NCPP

PLENARY MEMBERS

The plenary group was established to provide assistance and guidance to the task forces. The plenary was also responsible for discussing and ratifying recommendations on issues passed to it from the task forces, and assembling and approving the report of the project. The voting members of the group included all task force cochairmen, vice-cochairmen, and the Environmental and Industry Caucus chairmen.

Francis X. Quinn, S.J.
Plenary Chairman

Environmental Caucus
Laurence I. Moss, Chairman
John R. Bartlit
Paul P. Craig
Robert R. Curry
Norman L. Dean, Jr.
Ernst R. Habicht, Jr.
J. Michael McCloskey
Bruce J. Terris
Gregory A. Thomas
Grant P. Thompson

Industry Caucus
Gerald L. Decker, Chairman
Alan W. Beringsmith
Perry Brittain
Jackson B. Browning
John Corcoran
J. Robert Ferguson, Jr.
John H. Gibbons
Matthew Gould
Edwin R. Phelps
J. James Roosen

James T. B. Tripp

Edward V. Sherry

Macauley Whiting

MINING TASK FORCE

Environmental Members

J. Michael McCloskey, Cochairman, Sierra Club

Robert R. Curry, Vice Cochairman, University of Montana

Russell Boulding, Environmental Consultant (also served on Technical Staff)

Robert Golten, National Wildlife Federation

Norman Kilpatrick, Surface Mining Research Library

Arnold J. Silverman, University of Montana

Industry Members

John Corcoran, Cochairman, former chairman Consolidation Coal

Perry Brittain, Vice Cochairman, Texas Utilities

Edwin R. Phelps, Vice Cochairman, Peabody Coal Company

Gerald L. Barthauer, Consolidation Coal Company

Stanley Dempsey, AMAX, Inc.

Harrison Loesch, Peabody Coal Company

James Sembower, Dow Chemical, U.S.A.

Alten Grandt, Alternate, Peabody Coal Company

C. William Garrard, Alternate, Texas Utilities

David Itz, Alternate, Texas Utilities

Charles Kucera, Alternate, AMAX, Inc.

Joseph Mullan, Adviser, National Coal Association

Phillip Antommaria, Technical Staff, D'Appolonia Consulting Engineers, Inc.

TRANSPORTATION TASK FORCE

Environmental Members

James T. B. Tripp, Cochairman, Environmental Defense Fund
Gregory A. Thomas, Vice Cochairman, Sierra Club
W. Bruce Allen, University of Pennsylvania
Mohamed T. El-Ashry, Environmental Defense Fund
Patricia S. Record, Sierra Club
Bradley C. Gewehr, Technical Staff, Public Interest Economics Center

Industry Members

J. Robert Ferguson, Jr., Cochairman, United States Steel Corporation
Gerald Westbrook, Vice Cochairman, Dow Chemical Company
Robert Calhoun, Sullivan & Worcester
Les Moore, Gulf States Utilities
David Schwartz, Alternate, Sullivan and Worcester
William Zoller, Alternate, United States Steel Corporation

AIR POLLUTION TASK FORCE

Environmental Members

Bruce J. Terris, Cochairman, attorney-at-law
John R. Bartlit, Vice Cochairman, New Mexico Citizens for Clean Air and Water
Joseph J. Brecher, attorney-at-law
Michael D. Williams, John Muir Institute
Nathalie V. Black, Technical Staff, Law Offices of Bruce J. Terris
Suellen T. Keiner, Technical Staff Alternate, Law Offices of Bruce J. Terris

Industry Members

J. James Roosen, Cochairman, Detroit Edison Company
Jackson B. Browning, Vice Cochairman, Union Carbide Corporation
David M. Anderson, Bethlehem Steel Corporation
Thomas L. Montgomery, Tennessee Valley Authority
Jerrel D. Smith, Union Electric Company
Richard Porter, Bethlehem Steel Corporation

FUEL UTILIZATION AND CONSERVATION TASK FORCE

Environmental Members

Paul P. Craig, Cochairman, University of California, Berkeley

Norman L. Dean, Jr., Vice Cochairman, Environmental Law Institute

Clark Bullard, University of Illinois

Jerome Kohl, North Carolina State University

Robert Williams, Princeton

Susan J. Finnegan, Technical Staff, University of Texas

Industry Members

Matthew Gould, Cochairman, Georgia Pacific

John H. Gibbons, Vice Cochairman, University of Tennessee

Frank G. Feeley, Olin Corporation

Herbert D. Nash, Pennsylvania Power & Light Company

Byron R. Brown, Adviser, E. I. DuPont de Nemours and Company

James J. O'Connor, Adviser, *Power Magazine*

William B. Wilson, Adviser, General Electric Company

ENERGY PRICING TASK FORCE

Environmental Members

Grant P. Thompson, Cochairman, Environmental Law Institute

Ernst R. Habicht, Jr., Vice Cochairman, Environmental Defense Fund

Paul Joskow, Massachusetts Institute of Technology

Phillip Mause, Nathan, Mause & Thorpe

Industry Members

Alan W. Beringsmith, Cochairman Pacific Gas and Electric Company

Edward V. Sherry, Vice Cochairman, Air Products and Chemicals Corporation

Cliff Jensen, Union Carbide Corporation

George Petri, Monsanto Company

Larry L. Schedin, Northern States Power Company

Richard E. Weber, Public Service Electric and Gas Company

Jay Kennedy, Adviser, ELCON

AD HOC EMISSION CHARGE TASK FORCE

Environmental Members
Laurence I. Moss, Energy/Environment Consultant, former president, Sierra Club
Bruce J. Terris, attorney-at-law

Industry Members
Gerald L. Decker, Dow Chemical Company
Jackson B. Browning, Union Carbide Corporation
Larry L. Schedin, Northern States Power Company

PREVIOUS PROJECT PARTICIPANTS

Owing to other commitments, the following individuals withdrew from project participation before completion of the task force reports: Charles D. Baker, Harbridge House, Industry Chairman, Transportation Task Force; Barbara Brandon, University of Kentucky Law School, Environmental Vice Cochairman, Mining Task Force; Kathleen Hurson, Bechtel Corporation, Industry Member, Transportation Task Force; Leonard Lee Lane, Public Interest Economics Center, Environmental Chairman, Transportation Task Force.

Appendix B: Schedule of Meetings Held, Phase I

PLENARY GROUP

1977	January 18–19	Gaithersburg, Maryland
	April 5–6	CSIS, Washington, D.C.
	September 14–15	CSIS
	December 7–8	CSIS
1978	February 8–9	CSIS

MINING TASK FORCE

1977	February 24	Chicago, Illinois
	March 17	Big Brown, Texas
	May 17–18	Billings, Montana
	June 29–30	St. Louis, Missouri
	August 9–10	Pittsburgh, Pennsylvania, and several mines in Ohio and West Virginia
	September 29–30	Charleston, West Virginia
	October 20–21	CSIS
	November 10–11	CSIS

TRANSPORTATION TASK FORCE

1977	March 4	CSIS
	April 5	CSIS
	June 7	CSIS
	July 19–20	CSIS
	August 12	Boston, Massachusetts
	October 27–28	Pittsburgh, Pennsylvania

AIR POLLUTION TASK FORCE

1977	February 7	CSIS
	March 7	CSIS
	May 2–3	Pittsburgh, Pennsylvania
	May 31–June 1	CSIS
	July 18–19	CSIS
	September 26–27	CSIS
	October 31–November 1	Santa Fe, New Mexico

FUEL UTILIZATION AND CONSERVATION TASK FORCE

1977	March 16	CSIS
	May 17	San Francisco, California
	July 7	CSIS
	August 6	Denver, Colorado
	October 10	CSIS

ENERGY PRICING TASK FORCE

1977	February 10	CSIS
	March 18	CSIS
	June 7	CSIS
	June 29	CSIS

(Pricing Sub-Task Force: August 1–2)

	August 3–4	CSIS
	September 28–29	CSIS
	October 26–27	CSIS

AD HOC EMISSION CHARGE TASK FORCE

1977	October 18	CSIS

Appendix C: Contributors

The following foundations, government agencies, and corporations contributed to the financial support of the National Coal Policy Project:

FOUNDATIONS

Brown Foundation
Ford Foundation
Andrew W. Mellon Foundation
Rockefeller Foundation
Sarah Scaife Foundation
William and Flora Hewlett Foundation

GOVERNMENT AGENCIES

Tennessee Valley Authority
U.S. Department of Energy
U.S. Department of the Interior, Bureau of Mines
U.S. Energy Research and Development Administration (ERDA)
U.S. Environmental Protection Agency

CORPORATIONS

Air Products & Chemicals, Inc.
Allied Chemical Corporation
Allis-Chalmers Corporation
AMAX Coal Company
American Electric Power Co., Inc.
American Telephone & Telegraph
Atlantic Richfield Company
Baltimore Gas & Electric Company

BASF Wyandotte Corporation
Bethlehem Steel Corporation
Brown & Root, Inc.
Carolina Power & Light Company
Caterpillar Tractor Company
Celanese Corporation
Central Illinois Light Company
CF&I Steel Corporation
Champlin Petroleum Corporation
City of Colorado Springs
Columbus & Southern Ohio Electric Co.
Commonwealth Edison Company
Consumers Power Company
Continental Oil Company
Corning Glass Works
Dallas Power & Light Company
Dayton Power & Light Company
Detroit Edison Company
Dow Chemical Company
Dresser Industries
E.I. DuPont de Nemours & Company
Ebasco Services, Inc.
El Paso Natural Gas
Enserch Corporation
Esmark, Inc.
General Electric Company
General Public Utilities Corporation
Getty Oil Company
B. F. Goodrich Company
Gulf Oil Company
Halliburton Company
Hercules Incorporated
Houston Natural Gas
Interlake, Inc.
Jones & Laughlin Steel Corporation
Joy Manufacturing Company
Kaiser Aluminum and Chemicals Corporation
Kaiser Steel Corporation
Minnesota Power & Light Company
Monsanto Company
Montana-Dakota Utilities Company
Murcol, Inc.
Nevada Power Company
New York State Electric & Gas Corporation
Niagara Mohawk Power Corporation
North American Coal Corp.
Ohio Edison Company
Olin Corporation
Pacific Gas & Electric Company
Pennsylvania Power & Light Company
Philadelphia Electric Company
Phillips Petroleum Company
Potomac Electric Power Company
PPG Industries, Inc.
Public Service Electric & Gas Company
Public Service Indiana
Republic Steel Corporation
Rohm and Haas Company
Sabine Corporation
Sharon Steel Corporation
Southern Pacific Company
Stauffer Chemical Company
SUNOCO Energy Development Company

Tampa Electric
Texaco, Inc.
Texas Electric Company
Texas Electric Service
Texas Power & Light Company
Texas Utilities Services, Inc.
Toledo Edison Company
Transcontinental Gas Company
Union Carbide Corporation
Union Electric Company of St. Louis
United States Steel Corporation
Virginia Electric & Power Company
Westinghouse Electric Company
Wisconsin Electric Power Company
Wisconsin Public Service Company
H. B. Zachry, Company

Notes

The public record about the National Coal Policy Project is not extensive. Most large university libraries will have the major NCPP report: Francis X. Murray, ed., *Where We Agree: Report of the National Coal Policy Project*, 2 vols. (Boulder, Colo.: Westview Press, 1978). This two-volume report is summarized in 70 pages in a monograph that was widely distributed: National Coal Policy Project, *Where We Agree: Report of the National Coal Policy Project: Summary and Synthesis* (Washington, D.C.: Georgetown University, Center for Strategic and International Studies, 1978) (referred to hereafter as *Where We Agree, Summary*). A third widely available document, which reprints parts of *Where We Agree*, is a 347-page transcript of a hearing on the NCPP held by a House Committee: *National Coal Policy Project: Hearing before the Subcommittee on Energy and Power of the Committee on Interstate and Foreign Commerce, House of Representatives*, 95th Cong., 2d sess., April 10, 1978, Serial No. 95–138. This is referred to below as *National Coal Policy Project, House Hearing*. A second hearing included a general discussion of the NCPP: *Regulatory Negotiation: Joint Hearings before the Select Committee on Small Business and the Subcommittee on Oversight of Government Management of the Committee on Governmental Affairs, United States Senate*, 96th Cong., 2d sess., July 29-30, 1980, 1–48. This is referred to below as *Regulatory Negotiation: Joint Hearings, 1980*.

Other NCPP documents are hard to find. The most useful are J. Charles Curran, ed., *The National Coal Policy Project: A Report of a*

Seminar at the Colorado School of Mines, September 1978 (Washington, D.C.: Georgetown University Center for Strategic and International Studies, 1979); National Coal Policy Project, *The National Coal Policy Project: Final Report* (Washington, D.C.: Georgetown University, Center for Strategic and International Studies, 1981); Francis X. Murray and J. Charles Curran, *Why They Agreed: A Critique and Analysis of the National Coal Policy Project* (Washington, D.C.: Georgetown University, Center for Strategic and International Studies, 1982). A short article about the NCPP appeared in *Fortune* magazine: Tom Alexander, "A Promising Try at Environmental Detente for Coal," *Fortune*, February 13, 1978, 94–96, 100–102, but there is no extensive journalistic treatment of the NCPP beyond short news articles. Barbara Gray and Tina M. Hay have studied the NCPP from the standpoint of the social psychological theory of interorganizational relations: see Tina M. Hay and Barbara Gray, "The National Coal Policy Project: An Interactive Approach to Corporate Social Responsiveness," in Lee E. Preston, ed., *Research in Corporate Social Performance and Policy*, Vol. 7, 1985 (Greenwich, Conn.: JAI Press, 1985), 191–212; Barbara Gray, "Conditions Facilitating Interorganization Collaboration," *Human Relations* 38 (1985): 911–937; and Barbara Gray and Tina M. Hay, "Political Limits to Interorganizational Consensus and Change," *Journal of Applied Behavioral Science* 22 (1986): 95–112.

Data for this study were gathered during field work in Washington, D.C., in the fall of 1979 and in 1983 and 1984. Supported by a grant from the Government and Law division of the Ford Foundation, I interviewed forty people during my first stay there, including participants in the NCPP, executive-branch officials concerned with policies discussed by the NCPP, and interest-group executives and lobbyists. The modal interview was forty-five minutes; most of the interviews were conducted in person, although some were conducted by telephone. Subjects were encouraged to respond in an open-ended fashion about their views of the NCPP: their experiences if they were participants or their observations and opinions if they were executive-branch officials or lobbyists. In addition, I specifically asked for observations and opinions about the implementation process for the project's platform (see chapters 6, 7, and 8). Also during that stay, the NCPP was still meeting, and I was able to attend twelve hours of such meetings, as well as a two-and-a-half-day conference about the NCPP approach to conflict resolution.

As a recipient of a grant from the Russell Sage Foundation, I was able to continue my work in Washington during 1983 and 1984 and gained some perspective on the NCPP through informal discussions with people in the Washington policy-making community, particularly people interested in alternative modes of dispute resolution. At that time, I realized that the NCPP had the major effect of encouraging experimentation with regulatory negotiation, and I accordingly conducted eleven interviews with people interested in that concept. Since then I have followed the developing literature about this topic.

Data from the interviews added depth to my reading of the publications of the NCPP, congressional hearings about the project, and the project's "bible," *The Rule of Reason*. In addition, I received material from the NCPP's files on a request basis, including financial data and news clippings about the project.

CHAPTER 1. AN EXPERIMENT IN INTEREST GROUP THEORY

1. NCPP, *The National Coal Policy Project: Final Report* (Washington, D.C.: Georgetown University, Center for Strategic and International Studies, 1981), 20–21 (hereafter cited as *Final Report*).

2. These figures are based on my inspection of the records of the NCPP. No company gave more than 3.5 percent of the total NCPP budget.

3. NCPP, *Final Report*, 21.

4. See Francis X. Murray, ed., *Where We Agree: Report of the National Coal Policy Project*, 2 vols. (Boulder, Colo.: Westview Press, 1978).

5. *New York Times*, February 10, 1978, p. 1; Tom Alexander, "A Promising Try at Environmental Detente for Coal," *Fortune*, February 13, 1978, 94–96, 100–102.

6. Robert Stobaugh and Daniel Yergin, eds., *Energy Future* (New York: Random House, 1979), 106–7, and Sam H. Schurr, *Energy in America's Future* (Baltimore, Md.: Johns Hopkins University Press, 1979), 9–11, 540.

7. This observation is based on a number of conversations with scholars and visitors at the Resources for the Future Foundation when I was a guest scholar there, September 1979 through January 1980. I would like to thank that foundation for making its facilities available to me.

8. Philip J. Harter, "Negotiating Regulations: A Cure for Malaise," *Georgetown Law Journal* 71 (1982): 1–118; *Regulatory Negotiation: Joint Hearings before the Select Committee on Small Business and the Subcommittee on Oversight of Government Management of the Committee on Governmental Affairs, United States Senate*, 96th Cong., 2d sess., July 29–30, 1980, 1–48.

9. Gail Bingham, *Resolving Environmental Disputes: A Decade of Experience* (Washington, D.C.: Conservation Foundation, 1986), 17–18.

10. Louise Dunlap, director of the Environmental Protection Center, interview with author, September 1979.

11. Richard Ayres, litigation director of the Natural Resources Defense Council, interview with author, October 1979. For an academic statement of this point of view, see Douglas J. Amy, *The Politics of Environmental Mediation* (New York: Columbia University Press, 1987), esp. 111–112.

12. Frank Murray, director of the NCPP, interviews with author, September 1979.

13. Richard Ayres, interview with author, October 1979.

14. John O'Leary, interview with author, November 1979.

15. David Howard Davis, *Energy Politics*, 3d ed. (New York: St. Martin's, 1982), chap. 2, and Harry M. Caudill, *Night Comes to the Cumberlands* (Boston: Little, Brown, 1963).

16. My views on countervailing power are presented more extensively in Andrew S. McFarland, "Interest Groups and the Policymaking Process: Sources of Countervailing Power in America," in Mark P. Petracca, ed., *The Politics of Interests: Interest Groups Transformed* (Boulder, Colo.: Westview Press, 1992), 58–79.

17. Arthur F. Bentley, *The Process of Government* (Cambridge, Mass.: Belknap Press, 1967, orig. pub. 1908); David B. Truman, *The Governmental Process* (New York: Knopf, 1951); Earl Latham, "The Group Basis of Politics: Notes for a Theory," *American Political Science Review* 46 (June 1952): 376–397.

18. Robert A. Dahl, *Who Governs?* (New Haven, Conn.: Yale University Press, 1961); Charles E. Lindblom, "The Science of 'Muddling Through,'" *Public Administration Review* 19 (Spring 1959): 79–88; Nelson W. Polsby, *Community Power and Political Theory* (New Haven, Conn.: Yale University Press, 1963); Aaron B. Wildavsky, *The Politics of the Budgetary Process* (Boston: Little, Brown, 1964). My views on these writers are presented in Andrew S. McFarland, *Power and Leadership in Pluralist Systems* (Stanford, Calif.: Stanford University Press, 1969).

19. Dahl, *Who Governs?* 5.

20. This is discussed more extensively in Andrew S. McFarland, "Interest Groups and Theories of Power in America," *British Journal of Political Science* 17 (April 1987): 129–147.

21. Theodore J. Lowi, Jr., *The End of Liberalism*, rev. ed. (New York: Norton, 1979), and Mancur Olson, Jr., *The Logic of Collective Action* (Cambridge, Mass.: Harvard University Press, 1965).

22. Douglass Cater, *Power in Washington* (New York: Random House, 1964), and A. Lee Fritschler, *Smoking and Politics: Policy Making and the Federal Bureaucracy*, 4th ed. (Englewood Cliffs, N.J.: Prentice-Hall, 1989).

23. McFarland, "Interest Groups and Theories of Power in America."

24. Hugh Heclo, "Issue Networks and the Executive Establishment," in Anthony King, ed., *The New American Political System* (Washington, D.C.: Ameri-

can Enterprise Institute, 1978): 87–124; see the various authors collected in James Q. Wilson, *The Politics of Regulation* (New York: Basic Books, 1980); Thomas L. Gais, Mark A. Peterson, and Jack L. Walker, "Interest Groups, Iron Triangles, and Representative Institutions in American National Government," *British Journal of Political Science* 14 (1984): 161–185; Jeffrey M. Berry, *Feeding Hungry People: Rulemaking in the Food Stamp Program* (New Brunswick, N.J.: Rutgers University Press, 1984); William P. Browne, *Private Interests, Public Policy, and American Agriculture* (Lawrence: University Press of Kansas, 1988); Christopher J. Bosso, *Pesticides and Politics: The Life Cycle of a Public Issue* (Pittsburgh, Pa.: University of Pittsburgh Press, 1987).

25. John Stuart Mill, *Considerations on Representative Government*, (New York: Liberal Arts Press, 1958), 27–28.

26. Ibid.

27. Ibid., 54.

28. Ibid., 55.

29. Ibid., 54.

30. William K. Muir, Jr., *Legislature: California's School for Politics* (Chicago: University of Chicago Press, 1982).

31. Jane J. Mansbridge, *Beyond Adversary Democracy* (Chicago: University of Chicago Press, 1983).

32. See Paul J. Quirk, "The Cooperative Resolution of Policy Conflict," *American Political Science Review* 83 (1989): 905–922, and Lawrence Susskind, Lawrence Bacow, Michael Wheeler, eds., *Resolving Environmental Regulatory Disputes* (Cambridge, Mass.: Schenkman, 1983).

33. James Q. Wilson, *The Politics of Regulation* (New York: Basic Books, 1980).

34. Muir, *Legislature*.

35. Barbara Ferman, "The Politics of Exclusion: Political Organization and Economic Development" (Paper presented to the Urban Affairs Association Conference, March 1989, Baltimore, Md.).

36. Philippe C. Schmitter, "Still the Century of Corporatism?" *Review of Politics* 36 (1974): 85–131, and Peter J. Katzenstein, *Small States in World Markets* (Ithaca, N.Y.: Cornell University Press, 1985).

37. Katzenstein, *Small States*.

38. See, for example, Arthur F. P. Wassenberg, "Neo-Corporatism and the Quest for Control: The Cuckoo Game," in Gehrhard Lehmbruch and Philippe C. Schmitter, eds., *Patterns of Corporatist Policy-Making* (Beverly Hills, Calif.: Sage Publications, 1982): 83–108; Gudmund Hernes and Arne Selvik, "Local Corporatism," in Suzanne D. Berger, ed., *Organizing Interests in Western Europe* (New York: Cambridge University Press, 1981): 103–119; Alan Cawson, *Corporatism and Political Theory* (New York: Basil Blackwell, 1986): 106–125. Cawson, however, includes within his discussion of middle-level corporatism processes termed herein as subgovernmental politics, which in this book are distinguished both from corporatism and cooperative pluralism.

39. Susan E. Clarke, "Urban America, Inc.: Corporatist Convergence of Power in American Cities?" in Edward M. Bergman, ed., *Local Economies in*

Transition (Durham, N.C.: Duke University Press, 1986): 37–58; Peter Eisinger, "Do American States Do Industrial Policy?" *British Journal of Political Science* 20 (1990): 509–535; Susan B. Hansen, "Targeting in Economic Development: Comparative State Perspectives," *Publius* 19 (Spring 1989): 47–62, and Hansen, "Industrial Policy and Corporatism in the American States," *Governance* 2 (April 1989): 172–197; Virginia Gray and David Lowery, "Corporatist Foundations of State Industrial Policy," *Social Science Quarterly* 71 (1990): 3–24.

CHAPTER 2. COAL, STRIP-MINING, AND AIR POLLUTION: THE POLICY CONTEXT

1. David Vogel, *Fluctuating Fortunes: The Political Power of Business in America* (New York: Basic Books, 1989): chaps. 3, 4.
2. Andrew S. McFarland, *Public Interest Lobbies: Decision Making on Energy* (Washington, D.C.: American Enterprise Institute, 1976); and "Public Interest Lobbies Versus Minority Faction," in Allan J. Cigler and Burdett A. Loomis, eds., *Interest Group Politics* (Washington, D.C.: Congressional Quarterly Press, 1983): 324–353.
3. Jeffrey M. Berry, *Lobbying for the People* (Princeton: Princeton University Press, 1977), and Burton A. Weisbrod et. al., *Public Interest Law: An Economic and Institutional Analysis* (Berkeley: University of California Press, 1978).
4. Richard A. Liroff, *A National Policy for the Environment: NEPA and Its Aftermath* (Bloomington: Indiana University Press, 1976).
5. J. Clarence Davies III and Barbara Davies, *The Politics of Pollution*, 2d ed. (Indianapolis, Ind.: Bobbs-Merrill, 1975), and Charles O. Jones, *Clean Air: The Policies and Politics of Pollution Control* (Pittsburgh, Pa.: University of Pittsburgh Press, 1975).
6. Christopher J. Bosso, "Adaptation and Change in the Environmental Movement," in Allan J. Cigler and Burdett A. Loomis, eds., *Interest Group Politics*, 3d ed. (Washington, D.C.: Congressional Quarterly Press, 1991): 168–170.
7. See David Vogel, *Fluctuating Fortunes*, and Andrew S. McFarland, "Public Interest Lobbies Versus Minority Faction."
8. Robert Stobaugh and Daniel Yergin, eds., *Energy Future* (New York: Random House, 1979), and Congressional Quarterly, *Energy Policy*, 2d ed. (Washington, D.C.: Congressional Quarterly, 1981).
9. Andrew S. McFarland, *Public Interest Lobbies: Decision Making on Energy*.
10. *Congressional Quarterly Almanac, 1969* (Washington, D.C.: Congressional Quarterly, 1970): 735–746.
11. Sam H.Schurr, et al., *Energy in America's Future: The Choices Before Us* (Baltimore: Johns Hopkins University Press, 1979): 258–259, 263–264, 300–303.
12. Ibid., 28.
13. "Report by a Study Group Sponsored by the Ford Foundation and Admin-

istered by Resources for the Future," Hans H. Landsberg, chairman, in *Energy: The Next Twenty Years* (Cambridge, Mass.: Ballinger, 1974), 230–235.

14. Statements about business opinion are based on my participation during 1976 and 1977 in events sponsored by the Energy Policy Program of the American Enterprise Institute for Public Policy Research.

15. David Vogel, *Fluctuating Fortunes*.

16. James Q. Wilson, ed., *The Politics of Regulation* (New York: Basic Books, 1980): Chaps. 8, 10.

17. John E. Chubb, *Interest Groups and the Bureaucracy: The Politics of Energy* (Stanford, Calif.: Stanford University Press, 1983).

18. Andrew S. McFarland, "Energy Lobbies," in Jack M. Hollander and Harvey Brooks, eds., *Annual Review of Energy:* Vol. 9, *1984* (Palo Alto, Calif.: Annual Reviews, 1984): 501–528.

19. David Howard Davis, *Energy Politics*, 3d ed. (New York: St. Martin's Press, 1982), chap. 2; Walter A. Rosenbaum, *Energy, Politics and Public Policy* (Washington, D.C.: Congressional Quarterly Press, 1981); Richard A. Harris, *Coal Firms under the New Social Regulation* (Durham, N.C.: Duke University Press, 1985); Sam H. Schurr et al., *Energy in America's Future*, passim.

20. See note 10.

21. Harry M. Caudill, *Night Comes to the Cumberlands* (Boston: Little, Brown, 1963), and Francis X. Murray, ed., *Where We Agree: Report of the National Coal Policy Project*, vol. 2 (Boulder, Colo.: Westview Press, 1978): 111–161.

22. Stobaugh and Yergin, eds., *Energy Future*, 79–112, 136–182. This paragraph is also based on my impressions when I lived in Washington, D.C., for most of the period between 1974 and 1981. At that time, I had many conversations with staff members of environmental lobbies, and I participated in seminars with representatives of business at the American Enterprise Institute.

23. These issues were well reviewed in Schurr et al., *Energy in America's Future*.

24. See *Congressional Quarterly Almanac, 1972* (Washington, D.C.: Congressional Quarterly, 1973).

25. Walter A. Rosenbaum, *Environmental Politics and Policy* (Washington, D.C.: Congressional Quarterly Press, 1985): 246.

26. John Gaventa, *Power and Powerlessness: Quiescence and Rebellion in an Appalachian Valley* (Urbana: University of Illinois Press, 1980).

27. Walter A. Rosenbaum, *The Politics of Environmental Concern*, 2d ed. (New York: Praeger, 1977): 238–247, and C. Lenth, *Federal Conflict in Environmental Policy*: "The Regulation of Surface Coal Mining in Illinois and the Nation" (Ph.D. diss., University of Chicago, 1983).

28. Rosenbaum, *Politics of Environmental Concern*; E. E. Schattschneider, *The Semisovereign People* (New York: Holt, Rinehart and Winston, 1960); Grant McConnell, *Private Power and American Democracy* (New York: Knopf, 1966).

29. Rosenbaum, *Politics of Environmental Concern*, 239.

30. *Congressional Quarterly Almanac, 1973* (Washington, D.C.: Congressional Quarterly, 1974): 619.

31. Ibid., 615.
32. Rosenbaum, *Politics of Environmental Concern*, 237.
33. Stobaugh and Yergin, eds., *Energy Future*, 106.
34. John Gaventa, *Power and Powerlessness.*
35. Duane Chapman, *Energy Resources and Energy Corporations* (Ithaca, N.Y.: Cornell University Press, 1983): 195–199.
36. *Congressional Quarterly Almanac, 1973*, 617; hereafter, this reference source will be abbreviated *CQA*.
37. *CQA, 1974*, 767.
38. *CQA, 1974*, 765–767; *CQA, 1975*, 189.
39. *CQA, 1977*, 617–626.
40. *CQA, 1975*, 182, 185.
41. See note #36.
42. *CQA, 1972*, 932; *CQA, 1973*, 622.
43. *CQA, 1974*, 766–771; *CQA, 1973*, 619.
44. *CQA, 1974*, 771–773.
45. *CQA, 1975*, 773.
46. *CQA, 1975*, 189.
47. *CQA, 1975*, 189.
48. *CQA, 1975*, 177.
49. *CQA, 1976*, 127–128.
50. *CQA, 1977*, 623, 624, 626.
51. *CQA, 1977*, 626, and Walter A. Rosenbaum, *Energy, Politics, and Public Policy* (Washington, D.C.: Congressional Quarterly Press, 1981): 123.
52. Rosenbaum, *Environmental Politics and Policy*, 247–248.
53. Rosenbaum, *Politics of Environmental Concern*, chap. 5.
54. Ibid.; J. Clarence Davies and Barbara Davies, *The Politics of Pollution*; Charles O. Jones, *Clean Air*; Lettie McSpadden Wenner, *One Environment under Law* (Pacific Palisades, Calif.: Goodyear, 1976).
55. U.S. Congress, Office of Technology Assessment, *The Direct Use of Coal* (Washington, D.C.: U.S. Government Printing Office, 1978).
56. R. Shep Melnick, *Regulation and the Courts: The Case of the Clean Air Act* (Washington, D.C.: Brookings Institution, 1983), and Lettie McSpadden Wenner, *The Environmental Decade in Court* (Bloomington: Indiana University Press, 1982).
57. Melnick, *Regulation and the Courts*, chap. 5.
58. Rosenbaum, *Environmental Politics and Policy*, 130.
59. Ibid., 131.
60. Bruce A. Ackerman and William T. Hassler, *Clean Coal/Dirty Air* (New Haven, Conn.: Yale University Press, 1981).
61. Ibid., 36–37, 78.
62. Ibid., 97–100.
63. Martha Derthick and Paul J. Quirk, *The Politics of Deregulation* (Washington, D.C.: Brookings Institution, 1985).
64. This is my inference from the lack of mention of such lobbying in Acker-

man and Hassler's *Clean Coal/Dirty Air*, and such lobbying has not been mentioned to me in interviews. In spite of an apparent impact on support for President Ford's veto of SMCRA in 1975, the electric utility industry was normally regarded at that time as having less lobbying impact than the coal industry, railroads, or environmentalists in the coal policy area.

65. Melnick, *Regulation and the Courts*, chap. 4.

66. *CQA, 1977*, 633, 641, 642, 643, 645; *CQA, 1976*, 128–143.

67. *CQA, 1977*, 644, and Ackerman and Hassler, *Clean Coal/Dirty Air*, 36–37.

CHAPTER 3. AN ORGANIZING PRINCIPLE: THE RULE OF REASON

1. Daniel Yergin, "Conservation: The Key Energy Source," in Robert Stobaugh and Daniel Yergin, eds., *Energy Future* (New York: Random House, 1979): 165.

2. Ibid., 163–166.

3. J. Charles Curran, ed., *The National Coal Policy Project: A Report of a Seminar at the Colorado School of Mines, September 1978*, (Washington, D.C.: Georgetown University, Center for Strategic and International Studies, 1979), 9.

4. Ibid., 10.

5. Ibid.

6. Wessel, *The Rule of Reason*.

7. Ibid., viii.

8. This is an example of the type of organizational process described in Michael Cohen, James March, and Johan Olsen, "A Garbage Can Model of Organizational Choice," *Administrative Science Quarterly* 17 (March 1972): 1–25.

9. *Where We Agree: Summary*, 2–3.

10. Wessel, *The Rule of Reason*, 7.

11. Ibid.

12. Ibid., 8.

13. Ibid., 13–16.

14. Ibid., 21, 22.

15. Tina M. Hay and Barbara Gray, "The National Coal Policy Project: An Interactive Approach to Corporate Social Responsiveness," in Lee E. Preston, ed., *Research in Corporate Social Performance and Policy*, Vol. 7, *1985* (Greenwich, Conn.: JAI Press, 1985): 201–202.

16. Ibid., 203.

17. Wessel, *The Rule of Reason*, 23–24.

18. I. William Zartman and Maureen R. Berman, *The Practical Negotiator* (New Haven, Conn.: Yale University Press, 1982): 4, 5.

19. Thomas C. Schelling, *The Strategy of Conflict* (Cambridge, Mass.: Harvard University Press, 1961), and Howard Raiffa, *The Art and Science of Negotiation* (Cambridge, Mass.: Belknap Press, 1982).

20. Roger Fisher and William Ury, *Getting to Yes: Negotiating Agreement Without Giving In* (Boston: Houghton Mifflin, 1981).
21. Ibid., table of contents.
22. Wessel, *The Rule of Reason*, 137-140.
23. Fisher and Ury, *Getting to Yes*, 86.
24. Ibid., 87-88.
25. Ibid., 91.
26. Ibid., 92.
27. *Regulatory Negotiation: Joint Hearings 1980*, 14.
28. Francis X. Murray and J. Charles Curran, *Why They Agreed: A Critique and Analysis of the National Coal Policy Project* (Washington, D.C.: Georgetown University, Center for Strategic and International Studies, 1982), 1-2.
29. Ibid., 3.
30. NCPP, *Final Report*, 3.
31. Murray and Curran, *Why They Agreed*, vi.
32. NCPP, *Final Report*, 3.
33. *Where We Agree: Summary*, 8.
34. Murray and Curran, *Why They Agreed*, 2.

CHAPTER 4. EARLY HISTORY OF THE NATIONAL COAL POLICY PROJECT

1. J. Charles Curran, ed.,*The National Coal Policy Project: A Report of a Seminar at the Colorado School of Mines, September 1978* (Washington, D.C.: Georgetown University, Center for Strategic and International Studies, 1979), 10-11.
2. Ibid., 11.
3. Ibid., 11-12.
4. Francis X. Murray and J. Charles Curran, *Why They Agreed: A Critique and Analysis of the National Coal Policy Project* (Washington, D.C.: Georgetown University, Center for Strategic and International Studies, 1982), v.
5. Curran, *National Coal Policy Project*, 12.
6. Ibid.
7. Robert H. Salisbury, "An Exchange Theory of Interest Groups," *Midwest Journal of Political Science* 13 (February 1969): 1-32.
8. NCPP, *Final Report*, 1-2.
9. Ibid., p. 2.
10. Congressional Quarterly, *Energy Policy*, 2d ed. (Washington, D.C.: Congressional Quarterly, 1981): 106-108, 230, 232-233.
11. NCPP, *Final Report*, 2.
12. Murray and Curran, *Why They Agreed*, v-vi.
13. Ibid., vi.
14. Seymour Martin Lipset, *Political Man*, expanded ed. (Baltimore, Md.: Johns Hopkins University Press, 1981): 76, 104-105, 199-201.

15. John Gaventa, *Power and Powerlessness: Quiescence and Rebellion in an Appalachian Valley* (Urbana: University of Illinois Press, 1980).

16. Joseph P. Kalt and Mark A. Zupan, "The Politics and Economics of Senate Voting on Coal Strip Mining Policy: Inadequacies in the Economic Theory of Regulation," Discussion Paper Series E-82–10 (Cambridge, Mass.: John F. Kennedy School of Government, October 1982).

17. *Sierra Club v. Ruckelshaus*, 344 F. Supp. 254 (D.D.C.), *aff'd*, 4 Environ. Rep. Cas. 1815 (D.C. Cir. 1972), *aff'd by an equally divided court sub non.*

18. Bruce A. Ackerman and William T. Hassler, *Clean Coal/Dirty Air* (New Haven, Conn.: Yale University Press, 1981).

19. R. Shep Melnick, *Regulation and the Courts: The Case of the Clean Air Act* (Washington, D.C.: Brookings Institution, 1983).

20. *New York Times*, February 10, 1978, p. 1.

21. Evelyn H. Schlenker and Marc J. Jaeger, "Health Effects of Air Pollution Resulting from Coal Combustion," in A. E. S. Green, ed., *Coal Burning Issues* (Gainesville: University of Florida Press, 1980): 287.

22. Andrew S. McFarland, *Public Interest Lobbies: Decision Making on Energy* (Washington, D.C.: American Enterprise Institute, 1976): 67–77, 87–99.

23. *Where We Agree: Summary*, 4.

24. *Congressional Quarterly Almanac, 1983* (Washington, D.C.: Congressional Quarterly, 1984): 549–551.

25. John R. Bartlit, interview with author, October 1979.

26. Louise Dunlap, interview with author, September 1979.

27. Francis X. Murray, interview with author, May 1991.

28. Ibid.

29. *National Coal Policy Project, House Hearing*, 252.

30. Murray and Curran, *Why Whey Agreed*, 36.

CHAPTER 5. THE PROCESS OF CONSENSUS

1. *Sierra Club v. Ruckelshaus*, 344 F. Supp. 254 (D.D.C.), *aff'd*. 4 Environ. Rep. Cas. 1815 (D.C. Cir. 1972), *aff'd by an equally divided court sub non.*

2. Andrew S. McFarland, *Public Interest Lobbies: Decision Making on Energy*, (Washington, D.C.: American Enterprise Institute, 1976), 67–77, 89–99.

3. Martha Derthick and Paul J. Quirk, *The Politics of Deregulation* (Washington, D.C.: Brookings Institution, 1985), chap. 7.

4. Robert H. Wiebe, *The Search for Order, 1877–1920* (New York: Hill and Wang, 1967), 169–177.

5. *Where We Agree: Summary*, 17–23.

6. Theodore J. Lowi, *The End of Liberalism*, rev. ed. (New York: Norton, 1979), chap. 3.

7. *Where We Agree: Summary*, 65.

8. Richard F. Fenno, Jr., "The House Appropriations Committee as a Political System: The Problem of Integration," *American Political Science Review* 56

(June 1962): 310–324. See also Fenno, *The Power of the Purse*; *Appropriations Politics in Congress* (Boston: Little, Brown, 1966).

9. Aaron Wildavsky, *The Politics of the Budgetary Process*, 4th ed. (Boston: Little, Brown, 1984): v–xxxi, 222–280.

10. Ibid., 47–62.

CHAPTER 6. THE PLATFORM

1. This chapter presents a summary of the issues discussed within the NCPP and the agreements reached upon these issues. It is left to the reader to make judgments about the significance of the common interests so discovered.

2. See Robert Stobaugh and Daniel Yergin, eds., *Energy Future* (New York: Random House, 1979), 106–107; Sam H. Schurr et al., *Energy and America's Future* (Baltimore, Md.: Johns Hopkins Press, 1979), 11, 540; Charles Mohr, "Opposing Sides Agree on Shift of Oil to Coal," *New York Times*, February 10, 1978, p. 1; *Business Week*, February 13, 1978; Tom Alexander, "A Promising Try at Environmental Detente for Coal," *Fortune*, February 13, 1978, pp. 94–96, 100–102; President's Commission for a National Agenda for the Eighties, *Energy, Natural Resources, and the Environment in the Eighties* (Washington, D.C.: U.S. Government Printing Office, 1980), 42–43; U.S. Congress, Office of Technology Assessment, *The Direct Use of Coal* (Washington, D.C.: Government Printing Office, 1979), 179–180; League of Women Voters Education Fund, "Current Focus: Coal Use and Clean Air: Goals in Collision?" (League of Women Voters, 1978), 4.

3. Louise Dunlap, interview with author, September 1979; Richard Ayres, interview with author, September 1979; Douglas J. Amy, *The Politics of Environmental Mediation* (New York: Columbia University Press, 1987), 111–112.

4. Louise Dunlap, interview with author, September 1979.

5. *Where We Agree: Summary*, p. 26.

6. Allen V. Kneese and Charles L. Schultze, *Pollution, Prices, and Public Policy* (Washington, D.C.: Brookings Institution, 1975), 69–84, and Schultze, *The Private Use of Public Interest* (Washington, D.C.: Brookings Institution, 1977), 46–57.

7. U.S. Congress, House of Representatives, "The Federal Coal-Fired Power Plant Siting Act," introduced by Rep. Donald Pease, 97th Cong., 1st sess., H.R. 1430.

8. *Where We Agree: Summary*, 9–14.

9. Ibid., 9–10.

10. Ibid., 12–13.

11. Ibid., 11.

12. Ibid., 9.

13. Ibid., 12.

14. Ibid.

15. Ibid., iv, 12.

16. See note 7.
17. *Where We Agree: Summary*, 35-36, 37.
18. *Where We Agree*, vol. 2, 200.
19. Ibid., 201.
20. See John W. Kingdon, *Agendas, Alternatives, and Public Policies* (Boston: Little, Brown, 1984), 87–88, and Nelson W. Polsby, "Policy Initiation in the American Political System," in Irving L. Horowitz, ed., *The Use and Abuse of Social Science* (New Brunswick, N.J.: Transaction Books, 1971), 296–308.
21. Martha Derthick and Paul J. Quirk, *The Politics of Deregulation* (Washington, D.C.: Brookings Institutions, 1985), chap. 2 and passim.
22. Schultze, *Public Use of Private Interests*.
23. Steven Kelman, "Economists and the Environmental Muddle," *Public Interest* 64 (Summer 1981): 106–123.
24. Robert Cameron Mitchell, "Public Opinion on Environmental Issues: Results of a National Opinion Survey" (Washington, D.C.: Council on Environmental Quality, 1980) and "How 'Soft,' 'Deep,' or 'Left?' Present Constituencies in the Environmental Movement for Certain World Views," *Natural Resources Journal* 20 (1980), 345–358.
25. David Vogel, *Fluctuating Fortunes: The Political Power of Business in America* (New York: Basic Books, 1989), and Graham K. Wilson, *Business and Politics* (Chatham, N.J.: Chatham House, 1985), chap. 2.
26. Francis X. Murray, ed., *Where We Agree: Report of the National Coal Policy Project*, vol. 2 (Boulder, Colo.: Westview Press, 1978), 275.
27. Richard F. Fenno, Jr., "The House Appropriations Committee as a Political System: The Problem of Integration," *American Political Science Review* 56 (1962): 310–324; Aaron Wildavsky, *The Politics of the Budgetary Process*, 4th ed. (Boston: Little, Brown, 1984); David Braybrooke and Charles E. Lindblom, *A Strategy of Decision* (New York: Free Press, 1963).
28. These interviews were conducted by Frank Murray and J. Charles Curran of the NCPP's staff. Participants included Decker, Moss, John Corcoran, Robert Curry, and Grant Thompson.

CHAPTER 7. THE FAILURE OF LOBBYING

1. John Corcoran, telephone conversation with author, October 1979.
2. Tom Alexander, "A Promising Try at Environmental Detente for Coal," *Fortune*, February 13, 1978.
3. Charles Mohr, "Policy Panel Develops 200 Steps for Switching from Oil to Coal," *New York Times*, February 10, 1978, 1 et seq., and editorial page.
4. Ibid.
5. Robert Stobaugh and Daniel Yergin, eds., *Energy Future: Report of the Energy Project of the Harvard Business School* (Random House, 1979), and Sam H. Schurr et al., *Energy in America's Future: The Choices Before Us* (Baltimore, Md.: Johns Hopkins University Press, 1979).

6. Stobaugh and Yergin, eds., *Energy Future*, 106; emphasis added.
7. Ibid., 106–107; emphasis added.
8. Ibid., 107.
9. Schurr et al., *Energy and America's Future*, 9–10; emphasis added.
10. Ibid., 11.
11. Ibid., 540.
12. Ibid., 539.
13. Ibid., 540.
14. National Coal and Surface Mining Conference, Greenbrier, W. Va., November 19–21, 1979.
15. See Philip J. Harter, "Negotiating Regulations: A Cure for Malaise," *Georgetown Law Journal* 71 (October 1982): 38–40.
16. Andrew S. McFarland, *Common Cause: Lobbying in the Public Interest*, (Chatham, N.J.: Chatham House, 1984), 131–132.
17. Tom Alexander, "A Promising Try at Environmental Detente for Coal," *Fortune*, February 13, 1978, 102.
18. Francis X. Murray, ed., *Where We Agree: Report of the National Coal Policy Project*, vol. 1 (Boulder, Colo.: Westview Press, 1978), ix.
19. Ibid., 4.
20. Jeffrey M. Berry, *The Interest Group Society*, 2d ed. (Glenview, Ill.: Scott, Foresman, 1989), 164–172.
21. See "Coal Slurry Defeated," in *Congressional Quarterly Almanac, 1983* (Washington, D.C.: Congressional Quarterly, 1984), 549–551.
22. John Corcoran, interview with author, October 1979.
23. Andrew S. McFarland, *Public Interest Lobbies: Decision Making on Energy* (Washington, D.C.: American Enterprise Institute, 1976).
24. William A. Donohue, *The Politics of the American Civil Liberties Union* (New Brunswick, N.J.: Transaction Books, 1985), 256.
25. McFarland, *Common Cause*, 95–99.
26. McFarland, *Public Interest Lobbies*, chap. 5.
27. U.S. Congress, House of Representatives, "National Coal Policy Project," Hearings before the Subcommittee on Energy and Power of the Committee on Interstate and Foreign Commerce, 95th Cong., 2d sess., April 10, 1978, Serial No. 95-138; and U.S. Congress, Senate, "Regulatory Negotiation," Joint Hearings before the Select Committee on Small Business and the Subcommittee on Oversight of Government Management of the Committee on Government Affairs, 96th Cong., 2d sess., July 29 and 30, 1980.
28. Francis Murray and Ralph Nurnberger, interviews with author, September and October 1979. The opinions of Reps. Dingell and Ottinger are manifest in U.S. Congress, House, "National Coal Policy Project," 1–2, 109–110.
29. U.S. Congress, Senate, "Regulatory Negotiation."
30. Ralph Nurnberger, interviews with author, September and October 1979.
31. For the role of congressional staff at that time, see Michael J. Malbin, *Unelected Representatives: Congressional Staff and the Future of Representative Government* (New York: Basic Books, 1980), and Harrison W. Fox, Jr., and Su-

san Webb Hammond, *Congressional Staffs: The Invisible Force in American Lawmaking* (New York: Free Press, 1977).

32. Raymond A. Bauer, Ithiel de Sola Pool, and Lewis Anthony Dexter, *American Business and Public Policy* (New York: Atherton Press, 1968), 401–458.

33. John W. Kingdon, *Agendas, Alternatives, and Public Policies* (Boston: Little, Brown, 1984).

34. Walter A. Rosenbaum, *The Politics of Environmental Concern*, 2d ed. (New York: Praeger, 1977), 234–251.

35. *Congressional Quarterly Almanac, 1979* (Washington, D.C.: Congressional Quarterly, 1980), 683–684; this conflict of interest corresponds to the views of Grant McConnell, *Private Power and American Democracy* (New York: Knopf, 1966).

36. *Congressional Quarterly Almanac, 1979*, 683–684.

37. For lobbying rules for nonprofit organizations at that time, see "Public Charities Lobbying" and "Muskie-Conable Bill," in *Congressional Quarterly Almanac, 1976* (Washington, D.C.: Congressional Quarterly, 1977), 486–489.

38. Electric utilities in areas of increasing demand for electricity usually are neutral on the cogeneration issue or might even favor increasing cogeneration as a means to delay the construction of extraordinarily expensive new generating units in times of escalating interest charges for bond issues.

39. NCPP, *Final Report*, 98.

40. See *Federal Register*, 45:38 (February 25, 1980): 12214–35, and 45:56 (March 20, 1980), 17959–76.

41. NCPP, *Final Report*, 101.

42. Ibid., 102.

43. Schurr et al., *Energy in America's Future*, 159–176.

44. For a discussion of the 1977 strip-mining act and its impact on small mining companies, see Richard A. Harris, *Coal Firms under the New Social Regulation* (Durham, N.C.: Duke University Press, 1985), 84–93, 120–122.

45. *National Coal Policy Project, House Hearing.*

CHAPTER 8. THE NCPP AND REGULATORY NEGOTIATION

1. David M. Pritzker and Deborah S. Dalton, eds., Office of the Chairman, Administrative Conference of the United States, *Negotiated Rulemaking Sourcebook* (Washington, D.C.: U.S. Government Printing Office, January 1990).

2. Gail Bingham, *Resolving Environmental Disputes: A Decade of Experience* (Washington, D.C.: Conservation Foundation, 1986), 13–21.

3. Ibid., 121–123.

4. Ibid., xxiv.

5. Timothy J. Sullivan, *Resolving Development Disputes Through Negotiations* (New York: Plenum Press, 1984), 86–87.

6. Lawrence Susskind and Jeffrey Cruikshank, *Breaking the Impasse: Consen-*

sual Approaches to Resolving Public Disputes (New York: Basic Books, 1987), 130.

7. Philip J. Harter, "Regulatory Negotiation: The Experience So Far," *Resolve* (Winter 1984): 1. *Resolve* is "a quarterly newsletter on environmental dispute resolution," published by the Conservation Foundation, 1717 Massachusetts Ave., N.W., Washington, D.C., 20036.

8. For example, *Congressional Record*, August 4, 1983, remarks by Senator Levin, S11788, and *Congressional Record*, January 27, 1981, remarks by Rep. Pease, H202.

9. *Regulatory Negotiation: Joint Hearings, 1980*, 48–80.

10. Teresa M. Schwartz, "The Consumer Product Safety Commission: A Flawed Product of the Consumer Decade," *George Washington Law Review* 51 (November 1982): 57–73.

11. *Regulatory Negotiation: Joint Hearings, 1980.*

12. *1982 Report, Administrative Conference of the United States* (Washington, D.C.: Government Printing Office, 1983).

13. Philip J. Harter, "Negotiating Regulations: A Cure for Malaise," *Georgetown Law Journal* 71 (October 1982): 114–118, contains the ACUS recommendations about regulatory negotiation.

14. Ibid. An excellent 900-page compilation of materials regarding regulatory negotiation was published by the ACUS in 1990: Administrative Conference of the United States, Office of the Chairman, David M. Pritzker and Deborah S. Dalton, eds., *Negotiated Rulemaking Sourcebook* (Washington, D.C.: Government Printing Office, January 1990). This handbook contains a lengthy discussion of how to organize a regulatory negotiation process (negotiated rulemaking is a synonym for regulatory negotiation). It also contains a 500-page compilation of articles about the topic.

15. Henry H. Perritt, Jr., "Negotiated Rulemaking in Practice," *Journal of Policy Analysis and Management* 5 (1986), 482–495. Three other articles about regulatory negotiation by Perritt are reprinted in Pritzker and Dalton's *Negotiated Rulemaking Sourcebook.*

16. Perritt's "Negotiated Rulemaking" is a good summary of what is known about circumstances leading to successful regulatory negotiation. Discussions of this topic can be found in the *Negotiated Rulemaking Sourcebook* throughout, including other articles by Perritt.

17. *Federal Register*, 48:237 (December 8, 1983): 55076–98.

18. Lawrence Mosher, "EPA, Looking for Better Ways to Settle Rules Disputes, Tries Some Mediation," *National Journal* (March 5, 1983), 504–506.

19. Address delivered to the Second National Conference on Environmental Dispute Resolution, Marriott Hotel, Washington, D.C., October 1, 1984.

20. *Federal Register*, 50:44 (March 6, 1985): 9204–9230. See p. 9215 for the list of twenty-three participating organizations. The final rule was published in *Federal Register* 50:169 (August 30, 1985): 35374–35401.

21. Ibid., 50:67 (April 8, 1985): 13944–13959. The Defenders of Wildlife, the Audubon Society, and the National Wildlife Federation were the environmentalist

lobbies that participated in the emergency exemptions regulatory negotiation. The final rule was published in *Federal Register* 51:10 (January 15, 1986): 1896–1907.

22. Environmental Protection Agency, "The Environmental Protection Agency's Regulatory Negotiation Project" (Memorandum in the files of the EPA Regulatory Negotiation Project, Summer 1985).

23. *Federal Register* 50:138 (July 18, 1985): 29306–29322. See also Douglas B. Feavor, "FAA Finally Rules on Pilots' Rest Periods," *Washington Post*, July 19, 1985, A23.

24. Feavor, "FAA Rules on Rest Periods."

25. Perritt, "Negotiated Rulemaking in Practice."

26. Ibid., 489.

27. Ibid., 488.

28. Environmental Protection Agency, "Environmental Protection Agency's Regulatory Negotiation Project," 4.

29. Daniel J. Fiorino, "Regulatory Negotiation as a Policy Process," *Public Administration Review* 48 (1988): 771.

30. Daniel J. Fiorino, associate director, Policy Analysis, Environmental Protection Agency, interview with author, May 1991.

31. Pritzker and Dalton, eds., *Negotiated Rulemaking Sourcebook*, 336–343.

32. Ibid., 315–318.

33. NCPP, *Final Report*, 27.

34. 127 *Congressional Record*, January 27, 1981, H202.

35. 129 *Congressional Record*, August 4, 1983, S11788.

36. Harter, "Negotiating Regulations," 38.

37. See Pritzker and Dalton, eds., *Negotiated Rulemaking Sourcebook*, 346–347, 367.

38. Ibid., 345–366; 134 *Congressional Record*, n. 137, September 30, 1988, S13760–66.

39. Public Law 101-648, November 29, 1990; 104 *Stat.* 4969–4977.

40. 136 *Congressional Record*, n. 51, May 1, 1990, H1852–1855, 1857–1860, for the passage by the House.

41. For the Senate passage by voice vote see 135 *Congressional Record*, n. 109, August 4, 1989, S10060–64. For the Senate passage of the conference report see 136 *Congressional Record*, n. 128, October 4, 1990, S14580–83; for the House passage of the conference report see 136 *Congressional Record*, n. 145, October 22, 1990, H10966.

42. 136 *Congressional Record*, n. 51, May 1, 1990, H1855.

43. *Congressional Quarterly Almanac, 1990* (Washington, D.C.: Congressional Quarterly, 1991), 229–279.

44. Environmental Protection Agency, "OAR Discussion on the Use of Consultation and Consensus-Building Processes with External Interests for Implementing the Clean Air Act" (Memorandum, n.d.), 2. (OAR is an abbreviation for Office of Air and Radiation.)

45. Ibid., entire memorandum.

46. Environmental Protection Agency, "Agreement Reached on Clean Vehicle Fuels," press release, August 16, 1991. Also see Matthew L. Wald, "Environmental Negotiators Flesh Out Bare-Bones Law," *New York Times*, June 24, 1991; Terry Atlas, "9 Cities to Get Cleaner Car Fuel: Old Foes—Oil Firms, Environmental Groups—Sign Pact," *Chicago Tribune*, August 17, 1991; *New York Times*, (Midwest edition), "Cleaner-Burning Gas Planned for Some Cities," August 17, 1991.

47. This has been widely covered in the serious press. See, for instance, Michael Weisskopf, "Writing Laws Is One Thing—Writing Rules Is Something Else: The White House May Be Gutting the Clean Air Act," *Washington Post National Weekly Edition*, September 30, 1991–October 6, 1991. The final content of the EPA's reformulation of gasoline regulations is not clear as this book goes to press. One problem is that the EPA is in the process of issuing a series of different regulations dividing the subject of gasoline reformulation into standards, enforcement, and labeling of reformulated gasoline at the pump. Notices concerning gasoline reformulation regulations pertain to Title 40, Sec. 80 of the *Code of Federal Regulations*, where they are indexed in the *List of the CFR Sections Affected* (generally known as the LSA index). The *Federal Register* (57 [October 20, 1992]: 47771) footnotes eight previous regulatory notices concerning gasoline reformulation.

One major political issue delaying the issuance of gasoline reformulation regulations is the politics of gasohol. The regulatory negotiation process suggested strict limitations on the use of gasohol mixtures in the nine high-pollution cities, but in the summer of 1992 a powerful lobbying coalition of sixteen farm-state senators, the Archer-Daniels-Midland Company (largest refiner of ethanol for gasoline), and corn growers' associations (ethanol is derived from corn) lobbied to remove the restrictions upon gasohol in the gasoline reformulation regulations. In my judgment, this lobbying coalition is likely to be successful with the Clinton administration. See Michael Arndt, "Bush Wrestles with Ethanol Issue," *Chicago Tribune*, September 9, 1992.

48. Environmental Protection Agency, "Agreement Reached on Clean Vehicle Fuels," 2.

49. Ibid.

50. Ibid., attachment, "Reformulated Gasoline."

51. Ibid., attachment, "Environmental Protection Agency: Advisory Committee on Reformulated Gasoline, Antidumping, and Oxygenated Gasoline: Agreement in Principle."

52. Keith Schneider, "Utilities to Take Steps to Cut Haze at Grand Canyon," *New York Times* (Midwest edition), August 8, 1991, A1; Matthew L. Wald, "U.S. Agencies Use Negotiations To Pre-empt Lawsuits over Rules," *New York Times* (Midwest edition), September 23, 1991, 1.

53. Schneider, "Utilities to Take Steps to Cut Haze."

54. Environmental Protection Agency, "Agreement Calls for 90-percent Emissions Reduction for Power Plant near Grand Canyon," press release, August

8, 1991; Schneider, "Utilities to Take Steps to Cut Haze"; *Federal Register*; 56:156 (August 13, 1991): 38399–38404.

55. Schneider, "Utilities to Take Steps."

56. See Environmental Protection Agency, "Agreement Calls for 90-percent Emissions Reduction" and attachments: "Navajo Generating Station: Principal Elements of Agreement among SRP, EDF, and GCT"; "Arizona Visibility Implementation Plan, Memorandum of Understanding"; "Recommended Regulatory Requirements for the Navajo Generating Station Visibility Rulemaking."

57. *Federal Register* 56:192 (October 3, 1991): 50177.

58. Wald, "U.S. Agencies Use Negotiations."

59. Environmental Protection Agency, Office of Air and Radiation, "Proposed Acid Rain Rules: Overview," 400/1-91/038 (October 1991), 4.

60. Environmental Protection Agency, "Theme of the Month—February 1991: Talking Points: Building Consensus" (January 1992), 2.

61. Environmental Protection Agency, Office of Air and Radiation, "Proposed Acid Rain Rules: Overview"; "Permits: Proposed Acid Rain Rule," 400/1-91/035 (October 1991); "Allowance System: Proposed Acid Rain Rule," 400/1-91/034 (October 1991); "Excess Emissions: Proposed Acid Rain Rule," 400/1-91/037 (October 1991); "Continuous Emissions Monitoring: Proposed Acid Rain Rule," 400/1-91/036 (October 1991); "Auctions, Direct Sales, and Independent Power Producers Written Guarantee Regulations," 400/1-91/045 (December 1991).

62. Wald, "U.S. Agencies Use Negotiations." This is my impression based on telephone interviews with EPA officials in three different offices of that organization, June 1991 and February–March 1992.

CHAPTER 9. COOPERATIVE PLURALISM

1. See Christopher J. Bosso, *Pesticides and Politics: The Life Cycle of a Public Issue* (Pittsburgh, Pa.: University of Pittsburgh Press, 1987), 227–233, and Bosso, "Transforming Adversaries into Collaborators: Interest Groups and the Regulation of Chemical Pesticides," *Policy Sciences* 21 (1988): 3–22, esp. 19.

2. Bosso, *Pesticides and Politics*; Bosso, "Transforming Adversaries into Collaborators"; *Congressional Quarterly Almanac, 1986* (Washington, D.C.: Congressional Quarterly, 1987), 120; *Congressional Quarterly*, October 1, 1988, 2721.

3. Hans J. Morgenthau, *Politics Among Nations*, 3d ed. (New York: Knopf, 1964).

4. Thomas C. Schelling, *The Strategy of Conflict* (Cambridge, Mass.: Harvard University Press, 1960). See also Robert Axelrod, *The Evolution of Cooperation* (New York: Basic Books, 1984).

5. Paul J. Quirk, "The Cooperative Resolution of Policy Conflict," *American Political Science Review* 83 (1989), 905–922.

6. John Stuart Mill, *Considerations on Representative Government* (New York: Liberal Arts Press, 1958), chap. 7.

7. William K. Muir, Jr., *Legislature: California's School for Politics* (Chicago: University of Chicago Press, 1982).

8. Francis X. Murray and J. Charles Curran, *Why They Agreed: A Critique and Analysis of the National Coal Policy Project* (Washington, D.C.: Georgetown University, Center for Strategic and International Studies, 1982), 33.

9. Hugh Heclo, "Issue Networks and the Executive Establishment," in Anthony King, ed., *The New American Political System* (Washington, D.C.: American Enterprise Institute, 1978), 87–124, and Thomas L. Gais, Mark A. Peterson, and Jack L. Walker, "Interest Groups, Iron Triangles, and Representative Institutions in American National Government," *British Journal of Political Science* 14 (1984): 161–185.

10. Edward O. Laumann and David Knoke, *The Organizational State: Social Choice in National Policy Domains* (Madison: University of Wisconsin Press, 1987), chaps. 4, 5.

11. Jo Freeman, *The Politics of Women's Liberation* (New York: Longman, 1977; first published 1975).

12. Louise Dunlap, interview with author, September 17, 1979. Grant P. Thompson; interview with author, October 4, 1979.

13. Paul M. Sniderman, *A Question of Loyalty* (Berkeley: University of California Press, 1981).

14. For a bibliography about environmental mediation, see Gail Bingham, *Resolving Environmental Disputes: A Decade of Experience* (Washington, D.C.: Conservation Foundation, 1986), 257–274. The American Bar Association has sponsored research on alternative modes of dispute resolution since the mid-1970s. The Department of Urban Studies and Planning at MIT has been a major center for the study of environmental mediation and other cooperative negotiating strategies. Lawrence E. Susskind is probably the most influential writer of the MIT group (see Bingham bibliography). Other information is available from the National Institute for Dispute Resolution in Washington, D.C. Douglas J. Amy, *The Politics of Environmental Mediation* (New York: Columbia University Press, 1987), is an academic criticism of alternative dispute resolution experiments.

15. Charles L. Schultze, *The Public Use of Private Interests* (Washington, D.C.: Brookings Institution, 1977), and Martha Derthick and Paul J. Quirk, *The Politics of Deregulation* (Washington, D.C.: Brookings Institution, 1985).

16. Bruce A. Ackerman and William T. Hassler, *Clean Coal/Dirty Air* (New Haven, Conn.: Yale University Press, 1981), and Christopher J. Bosso, *Pesticides and Politics*.

17. President's Commission for a National Agenda for the Eighties, *The Electoral and Democratic Process in the Eighties* (Washington, D.C.: Government Printing Office, 1980); Lester C. Thurow, *The Zero-Sum Society* (New York: Basic Books, 1980); Ira C. Magaziner and Robert B. Reich, *Minding America's Business: The Decline and Rise of the American Economy* (New York: Harcourt Brace Jovanovich, 1982).

18. William K. Stevens, "Economy Tops Pennsylvania Agenda," *New York Times* (Midwest edition), January 21, 1987, p. 8.

19. Robert Cameron Mitchell, "Public Opinion and Environmental Politics in the 1970s and 1980s," in Norman J. Vig and Michael E. Kraft, eds., *Environmental Policy in the 1980s: Reagan's New Agenda* (Washington, D.C.: Congressional Quarterly Press, 1984), 51–74.

20. Laumann and Knoke, *The Organizational State*, and John P. Heinz, Edward O. Laumann, Robert L. Nelson, and Robert H. Salisbury, *The Hollow Core: Private Interests in National Policy Making* (Cambridge, Mass.: Harvard University Press, 1993).

21. Susan E. Clarke, "Urban American, Inc.: Corporatist Convergence of Power in American Cities?" in Edward M. Bergman, ed., *Local Economies in Transition* (Durham, N.C.: Duke University Press, 1986), 37–58.

22. Peter Eisinger, "Do the American States Do Industrial Policy?" *British Journal of Political Science* 20 (1990), 509–535; Susan B. Hansen, "Targeting in Economic Development: Comparative State Perspectives," *Publius* 19 (Spring 1989): 47–62, and Hansen, "Industrial Policy and Corporatism in the American States," *Governance* 2 (April 1989), 172–197; Virginia Gray and David Lowery, "Corporatist Foundations of State Industrial Policy," *Social Science Quarterly* 71 (1990), 3–24.

23. Walter A. Rosenbaum, *Environmental Politics and Policy* (Washington, D.C.: Congressional Quarterly Press, 1985), 119–120, 124–125.

References

Ackerman, Bruce A., and William T. Hasser. 1981. *Clean Coal/Dirty Air*. New Haven, Conn.: Yale University Press.

Alexander, Tom. 1978. "A Promising Try at Environmental Detente for Coal." *Fortune* 13 February, 94–96, 100–102.

Amy, Douglas J. 1987. *The Politics of Environmental Mediation*. New York: Columbia University Press.

Axelrod, Robert. 1984. *The Evolution of Cooperation*. New York: Basic Books.

Bauer, Raymond A., Ithiel de Sola Pool, and Lewis Anthony Dexter. 1968. *American Business and Public Policy*. New York: Atherton Press.

Bentley, Arthur F. 1967. *The Process of Government*. Cambridge, Mass.: Belknap Press of Harvard University Press.

_______. 1989. *The Interest Group Society*. 2d ed. Glenview, Ill.: Scott-Foresman.

_______. 1984. *Feeding Hungry People: Rulemaking in the Food Stamp Program*. New Brunswick, N.J.: Rutgers University Press.

_______. 1977. *Lobbying for the People*. Princeton: Princeton University Press.

Bingham, Gail. 1986. *Resolving Environmental Disputes: A Decade of Experience*. Washington, D.C.: Conservation Foundation.

Bosso, Christopher J. 1991. "Adaptation and Change in the Environmental Movement." In Allan J. Cigler and Burdett A. Loomis, eds., *Interest Group Politics*. Washington, D.C.: Congressional Quarterly Press.

_______. 1988. "Transforming Adversaries into Collaborators: Interest Groups and the Regulation of Chemical Pesticides." *Policy Sciences* 21: 3–22.

_______. 1987. *Pesticides and Politics: The Life Cycle of a Public Issue.* Pittsburgh, Pa.: University of Pittsburgh Press.

Braybrooke, David, and Charles E. Lindblom. 1963. *A Strategy of Decision*. New York: Free Press.

Browne, William P. 1988. *Private Interests, Public Policy, and American Agriculture*. Lawrence: University Press of Kansas.

Cater, Douglass. 1964. *Power in Washington*. New York: Random House.

Caudill, Harry M. 1963. *Night Comes to the Cumberlands*. Boston: Little, Brown.

Cawson, Alan. 1986. *Corporatism and Political Theory*. New York: Basil Blackwell.

Chapman, Duane. 1983. *Energy Resources and Energy Corporations*. Ithaca, N.Y.: Cornell University Press.

Chubb, John E. 1983. *Interest Groups and the Bureaucracy: The Politics of Energy*. Stanford, Calif.: Stanford University Press.

Clarke, Susan E. 1986. "Urban America, Inc.: Corporatist Convergence of Power in American Cities?" In Edward M. Bergman, ed., *Local Economies in Transition*. Durham, N.C.: Duke University Press.

Cohen, Michael, James March, and Johan Olsen. 1972. "A Garbage Can Model of Organizational Choice." *Administrative Science Quarterly* 17: 1–25.

Congressional Quarterly. 1981. *Energy Policy*. 2d ed. Washington, D.C.: Congressional Quarterly.

Costain, Anne N. 1980. "The Struggle for a National Women's Lobby: Organizing a Diffuse Interest." *Western Political Quarterly* 33: 476–491.

Curran, J. Charles, ed. 1979. *The National Coal Policy Project: A Report of a Seminar at the Colorado School of Mines*, September 1978. Washington, D.C.: Georgetown University, Center for Strategic and International Studies.

Davies, J. Clarence, III, and Barbara Davies. 1975. *The Politics of Pollution*. 2d ed. Indianapolis, Ind.: Bobbs-Merrill.

Dahl, Robert A. 1961. *Who Governs?* New Haven, Conn.: Yale University Press.

Davis, David Howard. 1982. *Energy Politics*, 3d ed. New York: St. Martin's Press.

Derthick, Martha, and Paul J. Quirk. 1985. *The Politics of Deregulation*. Washington, D.C.: Brookings Institution.

Donohue, William A. 1985. *The Politics of the American Civil Liberties Union*. New Brunswick, N.J.: Transaction Books.

Eisinger, Peter. 1990. "Do American States Do Industrial Policy?" *British Journal of Political Science* 20: 509–535.

Fenno, Richard F., Jr. 1966. *The Power of the Purse: Appropriations Politics in Congress*. Boston: Little, Brown.

———. 1962. "The House Appropriations Committee as a Political System: The Problem of Integration." *American Political Science Review* 56: 310–324.

Ferman, Barbara. 1989. "The Politics of Exclusion: Political Organization and Economic Development." Paper presented to the Urban Affairs Association Conference, March 1989, Baltimore, Md.

Fiorino, Daniel J. 1988. "Regulatory Negotiation as a Policy Process." *Public Administration Review* 48: 764–772.

Fisher, Roger, and William Ury. 1981. *Getting to Yes: Negotiating Agreement Without Giving In*. Boston: Houghton Mifflin.

Ford Foundation Study Group. 1979. "Report." *Energy: The Next Twenty Years.* Cambridge, Mass.: Ballinger.

Fox, Harrison W., Jr., and Susan Webb Hammond. 1977. *Congressional Staffs: The Invisible Force in American Lawmaking*. New York: Free Press.

Freeman, Jo. 1977. *The Politics of Women's Liberation*. New York: Longman.

Fritschler, A. Lee. 1989. *Smoking and Politics: Policy Making and the Federal Bureaucracy*. 4th ed. Englewood Cliffs, N.J.: Prentice-Hall.

Gais, Thomas L., Mark A. Peterson, and Jack L. Walker. 1984. "Interest Groups, Iron Triangles, and Representative Institutions in American National Government." *British Journal of Political Science* 14: 161–185.

Gaventa, John. 1980. *Power and Powerlessness: Quiescence and Rebellion in an Appalachian Valley*. Urbana: University of Illinois Press.

Gray, Virginia, and David Lowery. 1990. "Corporatist Foundations of State Industrial Policy." *Social Science Quarterly* 71: 3–24.

Hansen, Susan B. 1989a. "Industrial Policy and Corporatism in the American States." *Governance* 2: 172–197.

______. 1989b. "Targeting in Economic Development: Comparative State Perspectives." *Publius* 19: 47–62.

Harris, Richard A. 1985. *Coal Firms under the New Social Regulation*. Durham, N.C.: Duke University Press.

Harter, Philip J. 1984. "Regulatory Negotiation: The Experience So Far." *Resolve* (Winter): 1.

______. 1982. "Negotiating Regulations: A Cure for Malaise." *Georgetown Law Journal* 71: 1–118.

Hay, Tina M., and Barbara Gray. 1985. "The National Coal Policy Project: An Interactive Approach to Corporate Social Responsiveness." In Lee E. Preston, ed., *Research in Corporate Social Performance and Policy*, Vol. 7: 1985. Greenwich, Conn.: JAI Press.

Heclo, Hugh. 1978. "Issue Networks and the Executive Establishment." In Anthony King, ed. *The New American Political System*. Washington, D.C.: American Enterprise Institute.

Heinz, John P., Edward O. Laumann, Robert L. Nelson, and Robert H. Salisbury. 1993. *The Hollow Core: Private Interests in National Policy Making*. Cambridge, Mass.: Harvard University Press.

Hernes, Gudmund, and Arne Selvik. 1981. "Local Corporatism." In Suzanne D. Berger, ed., *Organizing Interests in Western Europe*. New York: Cambridge University Press.

Jones, Charles O. 1975. *Clean Air: The Policies and Politics of Pollution Control*. Pittsburgh, Pa.: University of Pittsburgh Press.

Kalt, Joseph P., and Mark A. Zupan. 1982. "The Politics and Economics of Senate Voting on Coal Strip Mining Policy: Inadequacies in the Economic Theory of Regulation," Discussion Paper Series E-82-10. Cambridge, Mass.: John F. Kennedy School of Government.

Katzenstein, Peter J. 1985. *Small States in World Markets*. Ithaca, N.Y.: Cornell University Press, 1985.

Kelman, Steven. 1981. "Economists and the Environmental Muddle." *Public Interest* 64: 106–123.

Kingdon, John W. 1984. *Agendas, Alternatives, and Public Policies*. Boston: Little, Brown.

Kneese, Allen V., and Charles L. Schultze. 1975. *Pollution, Prices, and Public Policy*. Washington, D.C.: Brookings Institution.

Ladd, Everett Carll. 1991. *The American Polity*, 4th ed. New York: Norton.

Latham, Earl. 1952. "The Group Basis of Politics: Notes for a Theory." *American Political Science Review* 46: 376–397.

Laumann, Edward O., and David Knoke. 1987. *The Organizational State: Social Choice in National Policy Domains*. Madison: University of Wisconsin Press.

Lenth, C. 1983. "Federal Conflict in Environmental Policy: The Regulation of Surface Coal Mining in Illinois and the Nation." Ph.D. dissertation, University of Chicago.

Lindblom, Charles E. 1959. "The Science of 'Muddling Through.'" *Public Administration Review* 19: 79–88.

Lipset, Seymour Martin. 1981. *Political Man*. Expanded ed. Baltimore, Md.: Johns Hopkins University Press.

Liroff, Richard A. 1976. *A National Policy for the Environment: NEPA and Its Aftermath*. Bloomington: Indiana University Press.

Lowi, Theodore J. 1979. *The End of Liberalism*. Rev. ed. New York: Norton.

McConnell, Grant. 1966. *Private Power and American Democracy*. New York: Knopf.

McFarland, Andrew S. 1992. "Interest Groups and the Policymaking Process: Sources of Countervailing Power in America." In Mark P. Petracca, ed., *The Politics of Interests: Interest Groups Transformed*. Boulder, Colo.: Westview Press.

______. 1987. "Interest Groups and Theories of Power in America." *British Journal of Political Science* 17: 129–147.

______. 1984. "Energy Lobbies." In Jack M. Hollander and Harvey Brooks, eds., *Annual Review of Energy*, Vol. 9, 1984. Palo Alto, Calif.: Annual Reviews.

______. 1983. "Public Interest Lobbies Versus Minority Faction." In Allan J. Cigler and Burdett A. Loomis, eds., *Interest Group Politics*. Washington, D.C.: Congressional Quarterly Press.

______. 1976. *Public Interest Lobbies: Decision Making on Energy*. Washington, D.C.: American Enterprise Institute.

______. 1969. *Power and Leadership in Pluralist Systems*. Stanford, Calif.: Stanford University Press.

Magaziner, Ira C., and Robert B. Reich. 1982. *Minding America's Business: The Decline and Rise of the American Economy*. New York: Harcourt Brace Jovanovich.

Malbin, Michael J. 1980. *Unelected Representatives: Congressional Staff and the Future of Representative Government*. New York: Basic Books.

Mansbridge, Jane J. 1983. *Beyond Adversary Democracy*. Chicago: University of Chicago Press.

Matthews, Donald R. 1960. *U.S. Senators and Their World*. Chapel Hill: University of North Carolina Press.

Melnick, R. Shep. 1983. *Regulation and the Courts: The Case of the Clean Air Act*. Washington, D.C.: Brookings Institution.

Mill, John Stuart. 1958. *Considerations on Representative Government*. New York: Liberal Arts Press.

Mitchell, Robert Cameron. 1984. "Public Opinion and Environmental Politics in the 1970s and 1980s." In Norman J. Vig and Michael E. Kraft, eds., *Environmental Policy in the 1980s: Reagan's New Agenda*. Washington, D.C.: Congressional Quarterly Press.

______. 1980a. "How 'Soft,' 'Deep,' or 'Left'? Present Constituencies in the Environmental Movement for Certain World Views." *Natural Resources Journal* 20: 345–358.

______. 1980b. "Public Opinion on Environmental Issues: Results of a National Opinion Survey." Washington, D.C.: Council on Environmental Quality.

Morgenthau, Hans J. 1964. *Politics Among Nations*. 3d ed. New York: Knopf.

Mosher, Lawrence. 1983. "EPA, Looking for Better Ways to Settle Rules Disputes, Tries Some Mediation." *National Journal* (March): 504–506.

Muir, William K., Jr. 1982. *Legislature: California's School for Politics*. Chicago: University of Chicago Press.

Murray, Francis X. 1978. *Where We Agree: Report of the National Coal Policy Project*, 2 vols. Boulder, Colo.: Westview Press.

Murray, Francis X., and J. Charles Curran. 1982. *Why They Agreed: A Critique and Analysis of the National Coal Policy Project*. Washington, D.C.: Georgetown University, Center for Strategic and International Studies.

Nadel, Mark V. 1971. *The Politics of Consumer Protection*. Indianapolis, Ind.: Bobbs-Merrill.

National Coal Policy Project. 1981. *The National Coal Policy Project: Final Report*. Washington, D.C.: Georgetown University, Center for Strategic and International Studies.

______. 1978. *Where We Agree: Report of the National Coal Policy Project: Summary and Synthesis*. Washington, D.C.: Georgetown University, Center for Strategic and International Studies.

Nivola, Pietro. 1980. "Energy Policy and the Congress: The Politics of the National Gas Policy Act of 1978." *Public Policy* 28: 491–543.

Olson, Mancur, Jr. 1965. *The Logic of Collective Action*. Cambridge, Mass.: Harvard University Press.

Ornstein, Norman J., and Shirley Elder. 1978. *Interest Groups, Lobbying, and Policymaking*. Washington, D.C.: Congressional Quarterly Press.

Perritt, Henry H., Jr. 1986. "Negotiated Rulemaking in Practice." *Journal of Policy Analysis and Management* 5: 482–495.

Polsby, Nelson W. 1971. "Policy Initiation in the American Political System." In

Irving L. Horowitz, ed., *The Use and Abuse of Social Science*. New Brunswick, N.J.: Transaction Books.

______. 1963. *Community Power and Political Theory*. 1st ed. New Haven, Conn.: Yale University Press.

President's Commission for a National Agenda for the Eighties. 1980a. *The Electoral and Democratic Process in the Eighties*. Washington, D.C.: U.S. Government Printing Office.

______. 1980b. *Energy, Natural Resources, and the Environment in the Eighties*. Washington, D.C.: U.S. Government Printing Office.

Pritzker, David M., and Deborah S. Dalton, eds. 1990. *Negotiated Rulemaking Sourcebook*. Washington, D.C.: U.S. Government Printing Office.

Quirk, Paul J. 1989. "The Cooperative Resolution of Policy Conflict." *American Political Science Review* 83: 905–922.

Raiffa, Howard. 1982. *The Art and Science of Negotiation*. Cambridge, Mass.: Belknap Press.

Rosenbaum, Walter A. 1985. *Environmental Politics and Policy*. Washington, D.C.: Congressional Quarterly Press.

______. 1981. *Energy, Politics, and Public Policy*. Washington, D.C.: Congressional Quarterly Press.

______. 1977. *The Politics of Environmental Concern*, 2d ed. New York: Praeger.

Salisbury, Robert H. 1969. "An Exchange Theory of Interest Groups." *Midwest Journal of Political Science* 13: 1–32.

Schattschneider, E. E. 1960. *The Semisovereign People*. New York: Holt, Rinehart and Winston.

Schelling, Thomas C. 1961. *The Strategy of Conflict*. Cambridge, Mass.: Harvard University Press.

Schlenker, Evelyn H., and Marc J. Jaeger. 1980. "Health Effects of Air Pollution Resulting from Coal Combustion." In A. E. S. Green, ed., *Coal Burning Issues*. Gainesville: University of Florida Press.

Schmitter, Philippe C. 1974. "Still the Century of Corporatism?" *Review of Politics* 36: 85–131.

Schultze, Charles L. 1977. *The Public Use of Private Interests*. Washington, D.C.: Brookings Institution.

Schurr, Sam, et al. 1979. *Energy in America's Future: The Choices Before Us*. Baltimore, Md.: Johns Hopkins University Press.

Schwartz, Teresa M. 1982. "The Consumer Product Safety Commission: A Flawed Product of the Consumer Decade." *The George Washington Law Review* 51: 57–73.

Sniderman, Paul M. 1981. *A Question of Loyalty*. Berkeley: University of California Press.

Stobaugh, Robert, and Daniel Yergin, eds. 1979. *Energy Future*. New York: Random House.

Stone, Clarence N. 1989. *Regime Politics: Governing Atlanta, 1946–1988*. Lawrence: University Press of Kansas.

Sullivan, Timothy J. 1984. *Resolving Development Disputes Through Negotiations*. New York: Plenum Press.

Susskind, Lawrence, and Jeffrey Cruikshank. 1987. *Breaking the Impasse: Consensual Approaches to Resolving Public Disputes*. New York: Basic Books.

Susskind, Lawrence, Lawrence Bacow, and Michael Wheeler, eds. 1983. *Resolving Environmental Regulatory Disputes*. Cambridge, Mass.: Schenkman.

Thurow, Lester C. 1980. *The Zero-Sum Society*. New York: Basic Books.

Truman, David B. 1951. *The Governmental Process*. New York: Knopf.

U.S. Congress. House. Committee on Interstate and Foreign Commerce. Subcommittee on Energy and Power. *National Coal Policy Project: Hearing before the Subcommittee on Energy and Power of the Committee on Interstate and Foreign Commerce, House of Representatives*. 95th Cong., 2d sess., April 10, 1978.

U.S. Congress. Office of Technology Assessment. 1979. *The Direct Use of Coal*. Washington, D.C.: U.S. Government Printing Office.

U.S. Congress. Senate. 1980. *Regulatory Negotiation: Joint Hearings before the Select Committee on Small Business and the Subcommittee on Oversight of Government Management of the Committee on Governmental Affairs*. 96th Cong., 2d sess., July 29–30, 1–48.

Vogel, David. 1989. *Fluctuating Fortunes: The Political Power of Business in America*. New York: Basic Books.

Wassenberg, Arthur F. P. 1982. "Neo-Corporatism and the Quest for Control: The Cuckoo Game." In Gehrhard Lehmbruch and Philippe C. Schmitter, eds., *Patterns of Corporatist Policy-Making*. Beverly Hills, Calif.: Sage Publications.

Weisbrod, Burton A., et al. 1978. *Public Interest Law: An Economic and Institutional Analysis*. Berkeley: University of California Press.

Wenner, Lettie McSpadden. 1982. *The Environmental Decade in Court*. Bloomington: Indiana University Press.

———. 1976. *One Environment Under Law*. Pacific Palisades, Calif.: Goodyear.

Wessel, Milton R. 1976. *The Rule of Reason: A New Approach to Corporate Litigation*. Reading, Mass.: Addison-Wesley.

Wiebe, Robert H. 1967. *The Search for Order, 1877–1920*. New York: Hill and Wang.

Wildavsky, Aaron. 1984. *The Politics of the Budgetary Process*. 4th ed. Boston: Little, Brown.

Wilson, Graham K. 1985. *Business and Politics*. Chatham, N.J.: Chatham House.

Wilson, James Q. 1973. *Political Organizations*. New York: Basic Books.

Wilson, James Q., ed. 1980. *The Politics of Regulation*. New York: Basic Books.

Yergin, Daniel. 1979. "Conservation: The Key Energy Source." In Robert Stobaugh and Daniel Yergin, eds., *Energy Future*. New York: Random House.

Zartman, I. William and Maureen R. Berman. 1982. *The Practical Negotiator*. New Haven, Conn.: Yale University Press.

Index

3 1542 00188 4083

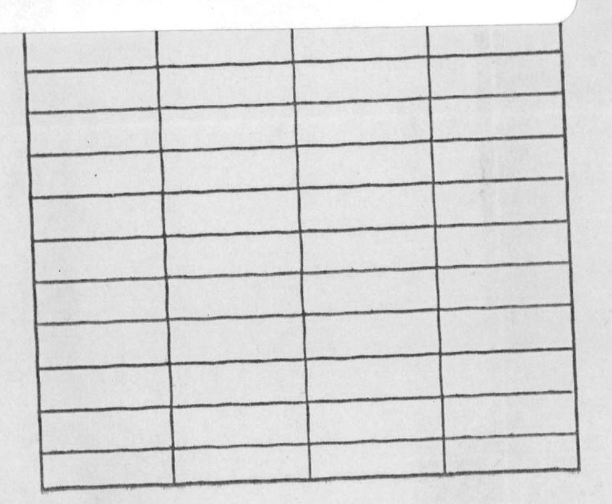